8138

WHERE WILL WE GO
WHEN THE SUN DIES?

WHERE WILL WE GO
WHEN THE SUN DIES?

JOHN W. MACVEY

STEIN AND DAY / Publishers / New York

First published in 1983
Copyright © 1983 by John W. Macvey
All rights reserved, Stein and Day, Incorporated
Designed by Louis A. Ditizio
Printed in the United States of America
STEIN AND DAY/*Publishers*/Scarborough House
Briarcliff Manor, N.Y. 10510

Library of Congress Cataloging in Publication Data

Macvey, John W.
 Where will we go when the sun dies?

 Bibliography: p.
 Includes index.
 1. Solar system. 2. Life on other planets.
3. Interplanetary voyages. I. Title.
QB502.M2   919.9′04   80-5386
ISBN 0-8128-2698-1

To those men and women of
Earth who will one day set out for the stars

Is it probable for Europe to be inhabited and not other parts of the World? Can one island have inhabitants and numerous other islands have none? Is it conceivable for one apple tree in the infinite orchard of the universe to bear fruit while innumerable other trees have nothing but foliage?

Konstantin Tsiolkovsky
Dreams of Earth and Sky, 1895

Acknowledgments

Once again it is my pleasant duty to acknowledge with gratitude all those who helped this book along the way. These include Miss Helen Campbell who once more has successfully transformed my ineffectual attempts at artistry into excellent diagrams and my daughter Karen who has had to cope for several months with the illegible scrawl which represents my handwriting.

I am also grateful to the British Interplanetary Society for details of their design for the star-ship *Daedalus*.

Last, and by no means least, I wish to express my sincere thanks to Benton Arnovitz, the editorial staff and all other members of Stein and Day who have at all times given their utmost encouragement, assistance, and advice.

To all these people, as much as to the author, this book is due.

Contents

$\mathcal{P}$*reface*

Which star lies nearest to the Earth? The conventional answer is Proxima Centauri, a dim red dwarf star which probably represents the third member of the Alpha Centauri system. This star, which is invisible to the naked eye, lies roughly 4.3 light years distant. Strictly speaking, however, the nearest star to our planet lies only eight light *minutes* from us. It is a much more familiar star and rather a bright one. We call it the Sun!

The Sun is a very average star in all respects. For this we can be very thankful. It is neither a giant nor a dwarf. With a surface temperature of approximately 6,000°C it is, by stellar standards, neither very hot nor very cool. And mercifully it is extremely stable—and likely to stay that way. Indeed it is only now reaching its prime. Another 5,000 million years must elapse before anything cataclysmic is likely to overtake it. At least so astrophysicists confidently believe, and in the present state of our knowledge we have no reason whatsoever to doubt them. On the other hand ideas change and thus, for all we know, so in time may our understanding of the life cycle of stars. Perhaps our familiar friendly Sun has an even longer life span before it than the 5,000 million years now projected. On the other hand its life expectancy could be less, just how much less would be the sheerest speculation but its end would probably still be sufficiently far ahead to preclude our losing any loss of sleep over the matter. Were the figure only a thousand million years, the consequences of the Sun's demise would still be of mere academic concern to us.

At the same time we must not be unmindful of the fact that some quite unexpected catastrophe could just conceivably overtake our central luminary. We know there are stars which explode catastrophically to become novae, some cataclysmically to become supernovae. Here too, men of science are reasonably certain this will not happen to the Sun. We repeat the words *reasonably certain* since total assurance in such a field is probably an impossibility. Were this by some strange and terrible streak of cosmic chance to be the destiny of our Sun it would not, of course, happen overnight. Presumably there would be a considerable period of warning, just how great we cannot say. It would probably be sufficiently long to give mankind a chance to do something about it. There would, understandably, be only one thing he *could* do and that is to head out for a safer planet surrounding another star. And this could be feasible only if his technology had reached the stage of rendering such a feat possible. Right now, of course, it is a mere pipe dream but one which, given long enough, may be developed and given practical expression.

The Sun has, as we are all aware, a system of nine planets swinging in orbit around it. A few lines back we described it as a very average star. It must then be asked whether the existence of a planetary system is also a prerequisite of an average star. For many years the Sun was regarded as an average star in all respects but one, that one being the possession of a planetary family. Today the pendulum of cosmological thought has swung in the opposite direction, to such an extent, in fact, that the existence of planetary systems is almost the accepted norm in respect of several categories of star. This is backed up by certain physical evidence in the case of a number of individual stars. At this point it might be reasonable to ask why we cannot give a more categorical answer.

Surely by turning our largest optical telescopes to these stars we should be able to ascertain directly whether or not they possess planets? Regrettably this is not possible; not, at least, from the surface of our world which, in effect, lies at the bottom of a great sea of turbulent air. Even the nearer stars are incredibly remote by terrestrial standards, so remote in fact, that even the greatest optical telescopes yet devised show them only as mere points of light lacking any discernible disc. Planets are very small bodies in comparison to their respective parent stars. Moreover they lie very close to them in comparison to the vast gulfs that separate star from star. Because of these inescapable factors it is impossible to establish directly the existence of "cousin" worlds.

Even a very large "super planet" would still be shrouded from our prying eyes by the intense glare of its relatively close and much larger parent star.

Thus the existence of extra-solar planets can be postulated only as a consequence of indirect mathematical analysis of a star's proper motion or by virtue of the fact that certain stars possess considerably less angular momentum than others. The newcomer to astronomy may feel these are rather hazy and nebulous reasons for accepting the existence of other planetary systems. In the ensuing pages however, we will endeavor to illustrate why, in fact, these reasons are soundly based.

Since it is as yet impossible to observe such incredibly remote worlds it might be thought, quite understandably, that there is not really very much we can say about them. This is becoming increasingly less so. We cannot as yet, of course, give chapter and verse concerning individual extra-solar planets. It is now possible, however, to dwell on many aspects of remote planetary systems. The whole concept of other solar systems is a very fascinating one, the more so since it is within some that other intelligences may one day be detected. As we will presently seek to show, there is little possibility of life in any form on the eight sister worlds which, in company with our own, sweep eternally around the Sun. In respect of our cousin worlds, the worlds of other stars, a very different state of affairs may often prevail.

The concept of other worlds, inhabited and otherwise, lying out in the great deeps of space is, by its very nature, highly intriguing. With the advent and continuing development of astronautics our own solar system has, in effect, shrunk. No longer is Mars the planet of mystery or Venus a tantalizing, cloud-enveloped sphere. They are now in our cosmic backyard. The frontiers of mystery and fascination have moved outward to the stars, to regions which at present no man or woman of Earth can hope to visit. The idea of other solar systems similar to, yet strangely different from our own, is just a part, but a very real part, of this exciting dimension. To that idea these pages are dedicated.

John W. Macvey
Saltcoats, Scotland
1983

WHERE WILL WE GO WHEN THE SUN DIES?

(1)

The Sun —
Our Lifeline

IN A SENSE the theme of this book is twofold. The universe is often said to be eternal and unchanging. This is not true, though so far as human lifetimes and, indeed, the whole of recorded history are concerned, no one would seriously question the validity of the statement. Astrophysicists, however, are showing an increasing tendency toward the belief that the entire universe is a pulsating entity. At present it is expanding and the evidence for this seems fairly irrefutable. The expansion should continue for another 35 thousand million years. At that point expansion is expected to cease and contraction to begin, a process which will probably last for another 55 thousand million years until once more all the matter within the universe is compressed into one super primeval "atom" (for want of a better word). The "fireball" effect postulated by Abbe G. Lemaitre and others will then be re-enacted and eventually another expanding universe will come into being. How long these cycles have been going on and for how long they will continue is hardly something we are ever likely to know. The best answer is probably throughout eternity and for all eternity.

The Sun is merely a single average star located about one third of the way from the rim of one very average galaxy. As we have just seen, the present universe, like all of us mortals, has a distinct and definite life span. There is, of course, something of a disparity between the time-scales! Galaxies have their lifetimes too. They are composed of thousands of millions of stars. Stars are born, they live, and then they die.

Our Sun, supremely vital to us, must also one day reach the end of the stellar road. But stars in dying go through some queer paroxysms. It cannot in all truth be said that death comes quickly, easily and uneventfully to them. For the most part they swell out, consume their nearer planets and roast their outer ones. Civilizations, great and small, good and bad, intelligent and stupid, are sponged from the scene with relentless and ruthless efficiency. Other stars flare up like great celestial beacons. We call such stars novae and admire their brilliance in our star-powered night sky. What this must mean to any planets they happen to possess and to any life on these planets is best left to the imagination.

It should not be imagined, however, that such dire stellar catastrophes occur overnight. In the case of stars dying a reasonably normal death (i.e., swelling out) the process takes many millions of years. In other words the peaceful yellow Sun we see in our skies in January will not become a red, planet-engulfing horror by December. It is almost certainly a fact that stars about to become novae give a period of warning, assuming of course there are astronomers on their worlds able to interpret the signs correctly. The time scale in this instance could be very much shorter than millions of years, with the final cataclysm very short indeed!

What does or what can a highly advanced technological civilization on the planet of a dying or potentially lethal star do? The answer is as stark as it is simple. It must get away from that star—or perish! If it is sufficiently developed, technologically the former is possible. If not, then its demise is surely at hand. In other words, occupants of planets orbiting suns of this kind must, while there is yet time, perfect a practical mode of interstellar travel and find themselves suitable worlds around "healthy" stars. There is no other way, nor can there ever really be.

Some day our Sun will almost certainly swell out to become a red giant star. In so doing Mercury, Venus, Earth, and Mars (probably the asteroids too) will assuredly be consumed—quite simply melted, then vaporized to become insignificant constituents of the Sun's outer atmosphere. In the Bible it is claimed that one day the Earth will be consumed by fire. Modern astrophysicists have no reason to dispute that belief! Jupiter (and probably Saturn too) will get a roasting and it is hard to predict how this will change them. To a lesser extent Uranus and Neptune will be affected too. Pluto, the small, enigmatic outer planet, may be rendered tropical or near tropical instead of the extreme frigid wilderness it must be today, but in the absence of a nitrogen-diluted oxygen atmosphere that planet would probably be of precious little use to us unless of course, the rise in temperature, being relatively gradual, generated an atmosphere appropriate to our needs. This seems highly improbable at best.

In the event, there would probably be no escape for the human race other than by mass migration to the stars, or more correctly to a suitable planet or planets of another star, assuming such a refuge exists and, of even greater importance, can be reached. In our first sentence we declared that the theme of this book was really twofold. This preamble relating to the first has thus brought us automatically to the second. To escape a solar holocaust there would have to be planets orbiting other stars. We believe such planets do exist. This aspect, its possibilities and promise (as well as the complications) will be examined at length. We must also have the technological means and expertise enabling us to reach some of them. Consequently we will also look briefly at forms of potential interstellar transit and the present state of the art. Thus our two themes merge to become one.

The necessity, for one reason or another, to escape from the Solar System is one which, understandably, has not gone unnoticed as an appropriate theme by writers of science-fiction. Many of these works are to be very highly commended. One that stands out prominently was published in 1939 or thereabouts. The increasingly urgent necessity to escape from the Solar System in this instance was not due to any actual or potential misdeeds on the part of the Sun. The menace lay in the fact that the part of the galaxy in which the Sun lies was gradually being obliterated. As I recall, all matter in this region of space was being transformed from mass into energy (the author presumably appreciated the implications of $E = mc^2$!). The result of all this was that, on a clear moonless starry night, a great patch of the sky was utterly devoid of stars. This region of blackness, of total "nothingness" was gradually reaching out toward the Sun and Solar System. Thus it had become imperative to find a planet or planets of a suitable star well away from the endangered region. Strangely enough most, if not all, astronomers failed to interpret the signs correctly. As a consequence the world went about its business quite oblivious to the mortal peril facing it in the foreseeable future. One did, but being considered a crank his warnings went totally unheeded. Indeed politicians, financiers and the like did their best to disgrace him, since his warnings might have an adverse effect on their interests. And so his desperate call for action, for immediate interstellar exploration and search for an alternative solar system were ignored or treated as an elaborate form of self-advertisement.

Unfortunately interstellar travel had gone undeveloped up till then. In such circumstances one would be tempted to regard the world as doomed and merely living on borrowed time, whether men of science appreciated the danger or not. However being science fiction, all (or nearly all) things are possible. This particular scientist and his young assistant (who also seemed to serve as his public relations agent) had perfected a hyper-light drive which rendered their small spacecraft capable of exceeding the speed of light many times over (here we see a glorious disregard for Einstein and the phenomenon of time dilation, etc.!). The drive worked (we'd have been surprised had it not!) and the two set off in an attempt to find a permanent safe haven for terrestrial civilization. Soon the expedition acquires a third member in the shape of a stowaway. As might be expected the stowaway is young, attractive—and decidedly female! The distraction which she offers to the scientist's assistant doesn't exactly appear to upset him! After many exciting adventures, a safe world orbiting a star almost identical to the Sun is found. The scientists of Earth eventually become convinced of the menace slowly enveloping their region of the galaxy. Migration is begun and, thanks to herculean efforts and organization, is completed in time—though only just? For good measure the young assistant and the stowaway are married, presumably to live happily ever after!

It seems highly improbable that this form of danger will ever menace the human race. It is to the center of the Solar System we must look—to the Sun itself. In chapter 8 (Uncertain Suns) we take a fairly close look at the life cycles of stars and how they vary. For now it is enough to say that our Sun should continue in its present state for some 5,000 million years. This is also roughly its age now, which renders it a benign and "healthy" middle-aged star. But after those 5,000 million years have elapsed, give or take a few million, the Sun, due largely to conversion of its hydrogen into helium, will become a bloated red giant star that consumes all its nearest planetary offspring, one of which is certain to be Earth.

In the circumstances this would hardly seem to be a cause for the loss of any sleep. After all, 5,000 million years *is* rather a long time. The threat of nuclear war, exhaustion of resources, pollution of the planet, not to mention escalating inflation, would all seem to represent a more valid reason for anxiety. Could the scientists be wrong? Could the Sun's life be measured not in millions of years but only in thousands?

It is a fact that the views of scientists can change in the light of increasing knowledge. It would not be the first time that theories and predictions have had to be abandoned, sometimes even reversed. (The process probably began with the "flat Earth" brigade.) At the present time however we must state quite categorically that there is *no* valid reason for suspecting that all our astrophysicists have got their sums wrong.

So far we have been looking at only one aspect. What about the Sun flaring up and becoming a nova? This prospect does seem to contain a greater element of risk. Yet, here too, modern astronomers are of the opinion that, so far as the Sun is concerned, such an event is unlikely. We hope they are right.

Novae are much more common than is popularly believed. Occasionally there is a very bright and spectacular one. It is these that make the headlines. In fact hardly a year passes without one or two faint, distant novae being detected by observatories. There are even a few dedicated amateur astronomers with the discovery of faint novae to their credit. About a hundred have been recorded during the last half century and it is certain that a still greater number must escape notice simply because they are missed in the stellar profusion appearing on photographic plates. The fact that most novae occur in the plane of the Milky Way only accentuates the difficulty. It has, as a consequence, been estimated that at least twenty to thirty novae occur in our galaxy annually. From this fact it would seem that we can draw two alternative conclusions. Either *all* the stars in the galaxy must have or will pass through a nova state, *or* only a certain type of star is capable of becoming a nova. In the second case it would then appear that the same star is given to showing nova characteristics at set intervals, some long, some short. It should probably be emphasized at this point that stars becoming novae are *not* utterly destroyed as is the case with the very much rarer and infinitely greater stellar explosions known as supernovae.

An increasing number of recurrent novae have been discovered and studied recently. A classic example, that of a star in the constellation Corona Borealis or Northern Crown, known to astronomers as T Coronae Borealis, was observed for the first time in 1866 only to reappear after eighty years in 1946. Its appearance on the second occasion, and the manner in which it fluctuated, matched almost exactly its performance in 1866. In similar fashion Nova T Sagittae in the constellation Sagitta, the Shield, first seen in 1913, also gave a repeat performance in 1946 (1946 seems to have been a good year for this sort of thing!). R.S. Ophiuchi also performed in this manner, in 1901 and 1903 respectively, though in this instance the outbursts were not exact replicas of one another. U Scorpii "went nova" on three occasions, 1862, 1906, and 1936. T. Pyxidis did the same in 1890, 1902, 1920, and (of course) *1946.*

The phenomenon of recurrence illustrates two cardinal points.
1. Only a very definite type of star acts in this manner.
2. Only the external layers of the star are expelled which, in effect, leaves the star relatively unscathed and able to repeat the performance at some later date.

It has also become apparent that the amplitudes of recurrent novae are less than those of normal nova which flare up only once. Whether of course, a star shows recurrent outbursts or flares up on one occasion can only be of academic interest to unfortunate peoples and civilizations esconced on a planet which orbits it closely. In this eventuality once is simply once too often.

In the past, novae were accounted for in ways which we know now are without a vestige of validity. An early supposition was that the outburst of heat and light was due to the mutual annihilation of two remote, faint stars coming into collision. In view of the tremendous distances separating star from star the chances of this are so remote as to be virtually nonexistent. Another concept which held sway for a time was the belief that the star concerned had passed through a cloud of diffuse interstellar matter. This is almost equally unlikely. There seems little doubt that the real origin of a nova lies *within* the star itself although the precise mechanism accounting for it is still a matter of considerable conjecture and debate.

The brightness of a star which has "gone nova" sometimes increases by as much as *ten* magnitudes. This means that the luminosity of the star involved has become about *10,000* times greater. Imagine, if you possibly can, our Sun quite suddenly turning up its power like this. Goodbye Earth and all the other inner planets! In the case of a supernova the luminosity increases to around *100 million* times the star's original intensity. If the Sun were to "go supernova" then it would be goodbye to the entire Solar System. By way of comparison, were Sirius to become a supernova, since it is only about eight light years distant, we here on Earth would be treated to an almost unbelievable spectacle, for we would receive as much light from it as we do from the full Moon! It would indeed be a marvellous and memorable sight, an intense and dazzling splash of light on the canvas of the night sky. Truly the landscape would be flooded in *starlight*.

Before we go into the subject of the Sun's ever becoming a nova let us try to imagine the horrifying predicament of a civilization facing such a dire catastrophe—one that, unfortunately, has *not* yet got around to establishing interstellar travel. Moreover the outer planets of their solar system to which they *could* escape are totally unsuitable. The sun in this system has already started to show signs of serious instability, the significance of which are all too clearly understood by astronomers and astrophysicists. Clearly there is nothing that can be done to prevent the impending outburst. Could they somehow protect their civilization or at least a part of it? Perhaps they might, by going underground to live a kind of troglodyte existence. This might work only if the outburst were of a reasonably minor character. Should the world and star of which we are speaking happen to be Earth and Sun respectively, such action would be of no avail since our planet, at a mere 93 million miles from the Sun, would simply be too close to escape vaporization. From a serious nova outburst, migration to a suitable planet of another star is the only real escape. Since a nova is generally an old star (and not a sprightly middle-aged one like the Sun) any planets surrounding it would also be old. Some may have been in existence long enough for the emergence of highly and technologically developed civilizations to have taken place. In that case the potential for mass migration *could* exist.

The first effects of a nova outburst on a planet such as our own would be terrible in the extreme. The sky remains a brilliant blue but only the molten or melting remains of great cities point to the recent existence of an old and cultured civilization. No longer is the Sun mild and benevolent. It has become a killer on the rampage. With a temperature that has now reached unprecedented and unbelievable heights even the sea has evaporated, its bed a crust of precipitated salts already running molten. The entire crust of the planet is riven and fractured by convulsive earth tremors and the whole sides of volcanoes are collapsing to release torrents of lava. The Earth, our once familiar world will soon be a cinder. Further rises in temperature will transform that cinder into mere superheated vapor. *Hic transit gloria mundi:* so passes the glory of the world.

Astronomers regard the transition of the Sun to a nova as highly unlikely. The evolution of a star slightly heavier than the Sun commences (as with the Sun) from a vast interstellar cloud of gas and dust which condenses under the influence of gravity to create a star and planets over a period of about 5,000 million years. The star then enters what is termed the "main sequence," i.e., it becomes a nice, placid, entirely normal, middle-aged sun. After about 5,000 million years it expands into a vast bloated star, referred to by virtue of its color as a "red giant." In so doing it consumes its inner planets, a fact upon which we have already touched. After a few thousand years of pulsating as a variable star it explodes as a nova, finally collapsing into a very small and extremely dense sort of star composed of degenerate matter known as a "white dwarf."

The evolution of a star with a mass akin to that of the Sun is probably very similar. From this it would appear that the nova state occurs *after* the red giant stage so that any planets (especially inner ones) have already been destroyed and the outer ones given their appointed roasting.

Astrophysicists must admit that we are still woefully ignorant on the subject of novae and supernovae. These phenomena may appear to be relatively rare events and indeed they are, if we use the average human lifetime as the criterion, but the state could well be the lot of most stars sooner or later. It is believed that all stars with a mass equal to or more than 1.44 times that of the Sun (a figure known as Chandrasekhar's Limit after the renowned astrophysicist of that name) must eventually become novae, casting off sufficient or rather superfluous mass to bring their masses within the safe limit just quoted. If this is so then the Earth is presumably safe. It is impossible to be certain however. With the increasing acquisition of fresh knowledge, ideas can change. It might become apparent, for example, that Chandrasekhar's Limit is set too high and that, far from being safe, our Sun just teeters uncertainly on the brink. Were this to be so then future generations might have to do some really fast thinking on the subject of interstellar travel, the location of suitable planets beyond the Solar System or, failing these things, the possibility of rendering the moons of our outer planets or even the planet Pluto itself somehow habitable. In the following chapter we will be having a brief, updated look at our sister worlds and their moons. This will be done for two extremely valid reasons:

1. A study of one solar system gives us ideas on the formation of solar systems in general and the form they may take.
2. Migration to the outer fringes of the Solar System might be the *only* alternative left to us if our Sun went on the rampage.

At this stage let us recap briefly. The Sun, in common with most stars, will eventually start to expand. As it swells out to become a typical red giant star it vaporizes all of its inner planets, Earth included. This should not happen for another five thousand million years so the danger can hardly be regarded as immediate. Our race, our whole planet as we know it, must have long changed by then. The menace represented by novae is perhaps less easily dismissed though astronomers now do not regard this as a cause for concern. Novae, recurrent novae and stellar variability are, in a general sense, all related, a subject we will be looking at later on, in chapter 8. Already our Sun shows a minor, roughly eleven-year period of variability and a sun-spot cycle of like period, the two being related. Any gradual increase in this variability, assuming that such a thing is possible, could in time render mandatory a quitting of the Solar System to avoid being alternately fried and frozen. Whether such action on the part of the Sun represents a prelude to the eventual assumption of nova status is, at present, the purest speculation.

One factor which may prove relevant in this respect is a recent claim that the Sun is *shrinking*. This claim was made recently by astronomer Jack Eddy of the Washington Observatory who has been examining measurements of the Sun's diameter made over the last two centuries. At the Royal Greenwich Observatory in England it is standard proceedure to record the times when the Sun's disc starts and finishes its transit of the zero meridian. In his examination of the Royal Observatory records going back over two centuries Eddy noticed that during this time there had been a consistent reduction in the time taken by the Sun to cross the meridian.

Now it is, of course, an accepted fact that the *apparent* size of the Sun's disc can vary with atmospheric conditions. Look, for example, how large the red setting sun over an ocean horizon appears in comparison with the golden yellow sphere at zenith position. This is an extreme example, and the real changes in the size of the Sun's disc referred to here are obviously only as a result of close scrutiny. It is, however, a source of error which undoubtedly introduces inaccuracies in measurements of the solar diameter from day to day, and might also be responsible for longer term trends in the apparent measured diameter of the Sun.

Eddy was able to eliminate this possibility by cross-checking data from the Royal Greenwich Observatory against measurements made at the Washington Observatory, the latter of course, being made under different environmental conditions. Records kept by the Washington Observatory over the last century indicated an overall reduction in the measured diameter of the Sun very similar to that reported by the Royal Greenwich Observatory. That the Sun has been shrinking over the last century, and almost certainly for very much longer, seems clearly indicated.

If we go back to the seventeenth century we find the celebrated astronomer Clavius recording a solar eclipse as annular (i.e., part of the Sun's disc remained in view around that of the Moon). According to the calculations of contemporary astronomers this eclipse ought to have been total (i.e., the Moon's disc coinciding exactly with that of the Sun). Assuming then that Clavius' observations are correct, the Sun's disc must at that time have been slightly too large to be completely hidden by that of the Moon.

Eddy's calculation would appear to indicate the Sun to be shrinking and at a fairly prodigious rate—something of the order of 1500 kms (or over 900 miles) per hundred years. It is perfectly obvious that a trend of this kind cannot possibly continue. If it did then the diameter in about 100,000 years would become zero and the Sun would simply cease to exist. Such an eventuality is clearly ridiculous. What this fact does point to, however, is that the Sun experiences pulsation over a very long period, just precisely how long is difficult to estimate. At the present point in the cycle the Sun would appear to be contracting. This, it seems reasonably safe to assume, will eventually cease and a long period of gradual expansion will begin. If this is so then the only conclusion that can be reached is that the Sun is indeed a variable star, but one having a very long period—period in terms of variability meaning the length of time between one maximum and the next.

At this time we can probably ask ourselves two questions, both of a rather obvious nature.

a. To what limit will this shrinkage extend?
b. When expansion starts, to what limits will *it* extend?

In other words, what we are saying is to what extent will our planet be frozen and then baked? Let us consider the first question. We may get a lead here because as we all know the Earth has experienced what are termed "Ice Ages." Thus if we try to measure the frequency and time scale of these we may just get a rough idea of the period of solar variability and the temperature range which this variability invokes.

To study the so-called ice ages we have to go back to what geologists term the Pleistocene epoch, sixth of the seven epochs that constitute the Cenozoic era of geological history. The Pleistocene was a period when glacial and interglacial conditions alternated over a large portion of the Earth's surface. During the glacial phases widespread continental ice sheets repeatedly covered large areas in the northern hemisphere, and glaciers were more numerous and extensive in both the northern and southern hemispheres. During the interglacial stages the climate seems to have been as warm or warmer than at present. Glaciated areas were reclothed with vegetation, soil was formed, and repopulation of animal life took place. The entire Pleistocene epoch is thought to have lasted for several hundreds of thousands of years.

Using the method of radiocarbon dating, ice ages would appear to have lasted from 10,000 to 55,000 years, those recording the lowest temperatures being about 15,000 years old. The Pleistocene began over two million years ago. The interglacial ages (i.e., warm periods) are thought to have been much longer than the glacial or ice ages. If the Sun is responsible for these variations by virtue of an inherent variability, then it would appear that the shrinkage phase and period of minimum diameter are shorter than the expansion phase and period of maximum diameter. Glaciations in the United States are named after the regions in which they took place. Thus we have the Illinoian, Kansan, and Nebraskan variations. The interglacial or warm periods for these regions are believed to have had periods of 100,000, 300,000 and 200,000 years respectively. If, therefore, we take an average for glacial duration as 30,000 years and for interglacial duration as 200,000 years we get a cycle of about 230,000 years. This would point to a period of solar variability of the order of 230,000 years. These it must be emphasized are very tentative figures and for the present no undue significance should be attached to them save that they *may* just indicate the period of solar variability.

At the present time we probably lie *between* ice ages and it may be that the apparently shrinking Sun is indicative of the advent of the next. Whether or not the coming of another ice age will cause our descendants to flee the Solar System is difficult to say. Technology will have increased but it may have increased only to a point enabling mankind to live amid a glaciated landscape. That after all might still be simpler to achieve than interstellar travel. Nevertheless the terrible effect of an ice age should not be underestimated. At the time of the last ice age the mountainous western part of North America was occupied by a vast complex of glaciers. Throughout the Canadian sector this formed an almost continuous blanket of ice. The area comprising the Atlantic to the Rockies, the northern part of the United States (as far south as New York City; Cincinnati, Ohio; St. Louis, Missouri; Kansas City; and Pierre, South Dakota) were buried beneath the Laurentian ice sheet. In New Hampshire it is reckoned to have been at least 5,000 feet thick. In other parts the figure may have been nearer 10,000 feet. Greenland and Iceland were completely ice-covered. About half of the European continent from the north coast of Norway to Kiev in southeast Europe was covered by the Scandinavian ice sheet which may also have attained a thickness of 10,000 feet.

Huge glaciers existed in Siberia and much of it was covered by the Siberian ice sheet. The southern hemisphere was even more completely enveloped in ice, much more so even than it is now. In the southern Andes glaciers spread westward to the ocean in Chile and eastward to the pampas of Argentina. Even with an advanced technology it is difficult to see how civilization could endure amid such conditions. The first threat from the Sun, ironically enough, may not be a case of too much light and heat but too little. But what, given long enough, occurs when the reverse process sets in, when the Sun begins to pulsate outwards again. Obviously the ice will melt and retreat as it must have done on many occasions before and the ocean levels will duly rise in turn. Temperate climates would become tropical, and those that are already tropical even more so. But we would not expect mankind to be seared from the face of the Earth.

If ice ages are the result of solar variability (and this is by no means proven) then our Sun can only be described as a very mildly variable star. This being so it is the contraction of the solar disc in the millenia to come that constitutes the great peril. But the variability pattern of a star is not, of necessity, something that is unalterable. We believe we are sure of our Sun's future but it would be a sanguine person who would say that we could ever be *truly* certain. The contractions might grow greater, the periods of expansion more protracted, the expansion itself of a greater extent.

Though we have clearly experienced a number of ice ages it does not appear as if the opposite extreme has ever really existed. The continuous survival of living organisms on this planet over the last three to four thousand million years bears reasonably eloquent testimony to that. But this refers only to continued existence and development. Whether or not a sophisticated civilization would wish to escape extremes of this nature if it had the means of escape is another matter entirely.

Among the brightest novae in relatively recent times might be included Nova Persei 1901, Nova Aquilae 1918 (almost as bright as Sirius), and Nova Puppis 1942. Around the remnant stars of these novae, gas clouds can still be observed expanding with velocities of hundreds of kilometers per second. Examination of the spectra of these clouds reveals the emission lines of hydrogen and of singly ionized oxygen atoms. Spectral changes from initial outburst to the final stages of a nova tend to be rather complex but in certain emission lines the elements hydrogen, helium, carbon, nitrogen, calcium, iron, titanium, scandium, chromium, strontium, and yttrium are present. Even 100 years after the advent of a nova faint emission lines can still be observed indicating the continuing emission of gases from the remnant stars with velocities of around 200 kms per second. It will be evident therefore that even an average ordinary nova represents a tremendous stellar cataclysm. The total energy emitted during a large nova is of the order of 10^{45} ergs. This, for the record, is equal to the total radiation from the Sun for *10,000 years.*

There exists a less spectacular danger inherent in the Sun but one possessing fewer dramatic effects. It could, however, represent a greater degree of urgency: whereas it would not involve the *death* or demise of the Sun it could, nonetheless, eventually render life on Earth either difficult or impossible. This involves what are generally termed solar flares. These occur in association with sunspots or on disturbed areas of the solar photosphere between sunspots. The photosphere, incidentally, is the second outer shell of the Sun, the outermost being known as the chromosphere though in fact there exists a thin layer where the two meet known as the reversing layer.

Solar flares form very quickly but their duration is usually of the order of a few minutes. Since they are closely associated with sunspots, which are phenomena with strong magnetic fields, they must presumably be regarded as some kind of electrical effect. They can be *extremely* bright on occasion and give rise to intense brief emissions of X-rays and ultraviolet light. Such cosmic rays as we receive from the Sun are probably also attributable to solar flares.

The light and other forms of electromagnetic radiation takes just over eight minutes to reach the Earth. The visible light is able to pierce our atmosphere but X-rays and ultraviolet radiation are almost completely absorbed by it. This leads to strong ionization in the lower layers of the ionosphere, which affects radio communication in a number of ways. Normal long distance shortwave transmissions (15-50 meters) simply fade out entirely while longer wave lengths suddenly find themselves endowed with a greatly extended range. The layer of the atmosphere which normally *reflects* short wave (i.e., high frequency) transmissions has been rendered "transport" to them so off they go into space for the benefit of Martians, Moon-girls, Venus-men or anyone else who happens to be around out there. The long wave (i.e., low radio frequency) signals are now *reflected* from an ionospheric layer that has become *opaque* to them so that they are *reflected* and heard in regions of the world where they would not normally be picked up. It need hardly be added that professional radio operators are not exactly over enthusiastic about solar flares because of the chaos they create.

About a day or so (normally about thirty hours) after the Earth has received this radiation, a fresh wave begins to arrive. It is corpuscular in form and comprised of electrons and protons. These are responsible for displays of the aurorae, magnetic storms, and other forms of disturbances in the ionosphere. In addition atomic nuclei are pumped out. These are much heavier than protons and travel with a higher velocity—almost 1,000 miles per second, in fact, so that they reach the Earth within an hour of the flare's occurring. These particles, apart from causing ionization in the atmosphere, also create secondary cosmic ray particles as well as gamma radiation, which is far more penetrating than X-rays. Earlier we mentioned that the Sun shows minor variability over an eleven-year period, the peak being characterized by a profusion of sunspots. Flares are consequently more intense and more frequent around this time. A number of exceedingly large sunspots which appeared on the face of the Sun on the morning of January 24, 1938 produced one of the most magnificent displays of the Aurora Borealis or "Northern Lights" which began around 7:00 P.M. on Tuesday, January 25. The timing was about right since some thirty hours had elapsed from the appearance of the sunspots.

There now seems little doubt that solar activity of this sort affects the meteorology of our planet. It is difficult for a variety of reasons to pin this down exactly, though duration and temperature of summers and winters as well as quantity of precipitation can probably be attributed to such solar activity. It has even been suggested that there could be a connection between it and the increased speed and extent of polar ice cap melting. It has also been suggested that an occasional increase in certain ailments and the number of suicides occurs at times of sunspot maxima. Secondary cosmic ray particles and gamma radiation, which reach a peak at this time, are also thought to be a cause of genetic change.

We can therefore be doubly thankful for the existence of our atmosphere. Not only does it provide the air we require for respiration, it also protects us from deadly radiation. But suppose with the passing of the centuries our atmosphere, for one reason or another, became gradually but perceptibly less protective. Already it has been suggested that a plethora of high altitude supersonic commercial jets could create an increasingly adverse effect. The continuing and growing use (read abuse) of countless millions of aerosol-type sprays might also in time prove harmful in this respect. In the general context these things are probably of fairly minor import. But suppose over two or three thousand years our atmospheric shield became increasingly pervious to such radiation. Or suppose that the frequency and especially the intensity of solar flares became greater or began to escalate so that the atmosphere which presently shields us was no longer able to do so. What then for civilization and the future of the fauna and flora on this planet? X-rays, cosmic rays, ultraviolet radiation and gamma rays, even in moderate excess, are positively harmful. The drenching of our civilization by them would be most serious and were the situation to go on escalating, the future for our kind would be extremely dismal. As we will see in the next chapter, the other planets of the Solar System are not exactly Utopias. Only the stars might provide a sanctuary before serious genetic changes and carcinogenic tumors on an immense scale rendered the people of this planet a dying race. In such circumstances we could not say that the Sun had died—merely that it was suffering from an incurable and contagious disease.

On one point we can be quite certain. The Sun will die neither now nor in the future after the fashion in which life dies. It will not as it grows older become dimmer until eventually it fades out. Neither will it "black-out" suddenly as does a man dying of cardiac arrest. In no sense is it, in the measurable future, going to become a dark, frigid sphere leaving its family of worlds, moons, and asteroids without warmth or light for all eternity.

The real danger to the Sun from our viewpoint, the one that might render it essential to quit the Solar System for new cosmic pastures, are the possibilities of increasing variability, of minor novae or escalating bouts of tremendous flare activity. These might only be mere unexpected "hiccups" in the life of a normal and thoroughly conventional star. But such "hiccups" could mean life or death to the human race.

(2)

Our Sister Worlds

BEFORE WE PROCEED to discuss the worlds of other suns, the "cousin" worlds of our own planet, we should perhaps devote a little time to an examination of Earth's sister worlds, i.e., those planets which, along with our own, orbit the Sun. Since the theme of this book is that of planets orbiting *other* stars this may seem an odd sort of beginning. On the other hand a clear understanding of our own Solar System may be of considerable assistance in the understanding of more remote ones. It is doubtful whether any of our sister worlds could represent a real refuge for humanity were our Sun to go on some kind of celestial rampage, though to an overcrowded, increasingly polluted and mineral-short Earth they might represent salvation. Certainly it would be easier and we presume cheaper to haul minerals from Mars than from a planet of Alpha Centauri, over four light years distant.

This book is entitled *Where Will We Go When the Sun Dies?* Our sister planets might be more appropriate in the context of "Where Will We Go When the *Earth* Dies?" To the layman, by now much more familiar with the idea of space travel, the answer to the second question might seem fairly straightforward—to the Moon, to Mars, to Venus, to Jupiter. This chapter, even if it achieves little more, should quickly and finally put an end to any such pleasantly naive beliefs.

It would be pointless in a book of this sort to go too deeply into the subject of our sister worlds when so much has already been written about them. However, in view of the vastly augmented amount of information now reaching us as a consequence of landing, orbiting and fly-by probes, many of the standard works on the solar planets are now hopelessly dated. To illustrate this point we need only consider works on Mars written a mere three to four decades ago. Incredibly, as it now seems, considerable time and space was devoted to such absorbing and intriguing subjects as the Martian canals and Martians. Not one writer even considered the likelihood of craters and vast active volcanoes on Mars. Many saw them as a possibility perhaps on Mercury. Craters, it seemed, were by and large the prerogative of the Moon.

The flow of information relating to the planets and moons of the Solar System can only be described as one of continuing escalation. Some fine volumes dealing with the Sun's family of worlds published around the turn of the century are now simply museum pieces. It is not even necessary to go back that far. On my desk is what was once considered a quite excellent little volume on the planet Mercury. It is a very slim one. It could hardly be otherwise since our knowledge of the closest planet to the Sun at the time it was published can only be described as negligible—and our lack of it abysmal. A map or chart purporting to represent the planet's surface had been painstakingly produced over the years 1924 to 1929 by one of the world's finest observers, using one of the best refracting telescopes in existence, yet all it showed was a collection of undefined dark and light areas. Somewhat incredibly, one of the dark regions had been named "Solitude Hermae Trismegisti." An astronomer in 1938, with the planet's close proximity to the Sun in mind, remarked that a more appropriate name would simply have been "Hell." An entire chapter was devoted to a serious discussion as to whether or not the planet might possess an *atmosphere*. Despite the fact that Mercury lay at its nearest only about sixty million miles from us, the sum total of our knowledge concerning the planet could be covered in less than a hundred pages—and much of that was erroneous. How times have changed.

In dealing with the worlds of *other* stars, our thoughts inevitably tend at times to dwell on the possibilities of alien life forms, especially those of high intelligence, representatives of which might one day reach the environs of the Sun. Fifty or sixty years ago similar thoughts were being entertained, even among some professional astronomers, of such possibilities within the Solar System itself. Martians were seen as a distinct possibility (after all they did have canals to sail around on!). Even Venusians (etymologically, "Venerians" is regarded as more accurate) were not wholly ruled out. Only the readers and writers of science fiction, however, were prepared to accept the existence of Mercurians, Jovians, Saturnians, Uranians, Neptunians—and, quite incredibly, Plutonians! Today few of us harbor any such illusions. Alas, even the Martians have left us, and with their passing science-fiction can surely never be the same again. This is not to suggest that Mars could *never* have had a civilization. Once representatives of our kind reach the surface of the legendary red planet it is just remotely possible that the artifacts of a once thriving society might be found somewhere beneath its shifting, wind-borne, red and ochre sands. But certainly the Mars we know today could never have been the Mars of H.G. Wells or the fabulous *Barsoom* of Edgar Rice Burroughs. (Progress, it seems, has its *dis*advantages, too.)

The fact that no other advanced life exists (at the moment we are not ruling out the possibility of very low-order life at some points within the Solar System) is due entirely to the fact that the other planets are quite simply unsuited to it. This fact must also be valid in respect of all other solar systems throughout the entire universe. It seems reasonable to assume that, in most only one or two planets will be biologically appropriate, and, no doubt in some, none at all. Other planetary systems do not therefore *automatically* mean alien life forms, advanced, intelligent, moronic, or otherwise. This cardinal fact should always be borne in mind. Our own planetary system only narrowly escaped being barren. But for the existence of Earth it almost certainly would be. The nature of a star, its age, and planetary distance from it must apply out there in the depths of space just as much as it does here in the environs of the Sun.

In this very brief updated examination we will follow the normal practice of commencing with the innermost member of the Solar System and work outwards.

Mercury: Were the Sun to become hotter or to drench the Earth with harmful radition, Mercury is certainly the *last* planet to which we would flee. But Earth in the grip of a new ice age brought about by solar contraction might make it seem a potential haven. In no circumstances, however, could this planet be regarded as a sanctuary. It is a small world, with a diameter less than 3,000 miles. It is therefore not much larger than the Moon. Because of its close proximity to the Sun (about 33 million miles) and its low escape velocity, there is little possibility of any atmosphere. Prior to the advent of the space age Mercury was often referred to as "the elusive planet," being seen close to the Sun and setting just before it in the evening or rising just before the Sun in the mornings, depending on the time of year. It tends therefore to be a difficult body to study from the Earth. Because of this handicap Mercury tended to be neglected. The standard map of the planet's surface had been made by E. M. Antoniadi during the late twenties. To produce this Antoniadi had used the 33-inch refracting telescope of the Meudon Observatory, near Paris. The most he could achieve however, was a map showing dark patches and a number of brighter ones, which in fact meant very little. Even these turned out to be inaccurate. Indeed the map might as well not have been made. Only in 1974 when the American *Mariner 10* approached the planet did we learn anything definite about the surface features of Mercury.

Due to the fact that it lies so close to the Sun (the distance ranges from 29 to 43 million miles) Mercury experiences great extremes of temperature, but at all times it must be very, very hot indeed. Its year, or orbital period around the Sun, corresponds to 88 terrestrial days, and a rotation time on its axis is equivalent to 58.5 terrestrial days. Due to its eccentric orbit the planet's orbital speed must vary. In conjunction with its slow axial rotation the Sun would move erratically across the Mercurian sky.

The maximum surface temperature of the planet is of the order of 700°C. However during the long Mercurian night the temperature must plummet rapidly downwards, owing to the lack of atmosphere which, were it present, would act as a buffer, thereby retaining some of the heat after the Sun had set. Until the advent of instrumented space probes it was widely accepted that Mercury's "day" and "year" were of equal duration and equivalent to

88 terrestrial days. This situation arose from the belief that Mercury kept the same face permanently turned toward the Sun much as the Moon keeps one face permanently turned toward the Earth. (Until 1959 no one had the faintest idea what the far side of the Moon looked like.) Had this really turned out to be the case, Mercury would have had a hemisphere of permanent day and one of permanent night. Admittedly there would probably have been a narrow sort of twilight zone where "day" and "night" hemispheres met and over which the Sun would have moved alternately just above and just below the horizon. During the 1960s however this concept, for long so popular, was disproved as a result of radar observations. From the point of view of manned landings this is perhaps unfortunate since the existence of a "twilight" zone would have meant that at least part of Mercury enjoyed an equitable temperature. Any such comforting illusions must now be shed. Astronauts landing on Mercury (if ever) are going to find it alternately very hot and very cold, very light and very dark.

A hasty glance at the most recent close-up photographs of Mercury can very easily cause it to be confused with the Moon. This is entirely due to the great profusion of lunar-like craters. Closer scrutiny does, however, reveal one very considerable difference. On Mercury there are very few of the lunar-type maria or "seas," those immense plains which are as much a feature of the Moon as are its craters. The existence of Mercurian craters are assumed to be due to the same process as those that gave rise to them upon the Moon, since they appear to observe the same laws of distribution Craters having radial ray systems also occur on Mercury, another feature which this little planet has in common with the Moon.

At one time it was believed that, due to the tremendously high temperatures, lead from mineral veins in the rocks might have melted to form little pools fed by rivulets of molten metal. Zinc and tin were even envisaged as perhaps lying around in a near molten or at least plastic state. Present evidence, however, strongly suggests that the crust of Mercury is closely akin to that of the Moon on which, of course, no rivers or lakes of molten lead have been found nor are ever likely to be.

Astronauts setting foot on Mercury are almost certainly going to find conditions much more trying, extreme and dangerous than those on the Moon. We can never envisage domed terrestrial colonies on this stark, baked cinder of a world. Better somehow to try and exist in an Earth ice-age. There can be little doubt however, that automatic probes landed on its surface would prove of considerable value because of their potential for monitoring the Sun from a relatively close range.

It is not very difficult to imagine conditions on the surface of Mercury during the long period of its "day"—a desolation of craters, thermally degraded rock and probably masses of igneous rock which have flowed or been extruded from the planet's crust. In the absence of an atmosphere the silence would be as total and absolute as the terrifying loneliness. And there, hanging in the sky, a dreadful Sun, three times as large as when seen from Earth. And that same Sun is pouring a withering torrent of heat, light, and harmful radiation onto the planet's racked and tortured surface. Come what may, no salvation is to be found on Mercury.

Venus: Outward from Mercury for about 30 million miles we come to the planet Venus. It might be hard to imagine there could exist a planetary surface of more nightmarish nature than that of Mercury but so far as Venus is concerned, this is no less than the truth. What makes the situation bizarre is the fact that Venus, named after the legendary goddess of love, is such an exquisitely brilliant and beautiful object in the skies of Earth, whether as the "morning star" rising shortly before the Sun or the "evening star" hanging like a celestial beacon in the afterglow following the setting of the Sun. In the past Venus was frequently referred to as Earth's twin, since in respect to size and mass the two planets are indeed strikingly similar. Thereafter all resemblance between the two vanishes, as we will now show.

Seen through a telescope, even one of considerable aperture, Venus, despite its great brilliance, has always proved a frustrating and disappointing object showing nothing more than phases similar to those of the Moon and a very bright, blank disc. Occasionally transient features made their appearance but these were always ill-defined and generally very vague. This is hardly surprising, for the surface of the planet is permanently shrouded by a dense, opaque atmosphere. Thus in the pre-space era there was considerable irony in the fact that Venus, closest celestial object to us after the Moon, remained shrouded in total mystery.

So far as the nature of the surface was concerned, theories proliferated. There was the ocean-covered world in which some of the extensive carbon dioxide in the atmosphere had dissolved under pressure, giving a kind of "soda-water" sea; there was a replica of Earth during the Carboniferous period when the coal measures were being laid down, and there was the petroleum "sea" largely composed of hydrocarbons. Few envisaged a rocky desert—a pity, for they would have been right.

All the popular concepts regarding Venus were quite rudely shattered in 1962 when the first successful planetary probe, *Mariner 2,* flew by the planet. In so doing it was able to collect and transmit back to Earth a fair amount of valuable and, to some extent, totally unexpected information. For a start the surface temperature proved very much higher than had been anticipated. The first casualty, therefore, was the attractive "soda-water" sea theory, quickly followed by the other two.

The rotational period or "day," hitherto a matter for considerable conjecture, was found to be approximately 243 terrestrial days. To complicate matters further, the Venusian year turned out to be some 225 terrestrial days. The "day" on Venus is thus longer than the "year"! And, as if all this were not enough, it was also found that Venus possessed a *retrograde* rotation, i.e., instead of revolving on its axis from west to east like Earth, it goes the other way round. Already it was becoming plain that Venus was very far indeed from being the twin of Earth.

Following the highly successful *Mariner 2,* radar research carried out in the United States enabled the first crude map of the planet's surface to be produced. This indicated the presence of craters here also, though of a more shallow type than on the Moon or on Mercury. The true nature of the planet's surface was further revealed during December 1970, when the Russian probe, *Venera 7,* "soft-landed" on Venus. Fortuitously the equipment remained functional for an hour before succumbing to the extremely rigorous conditions prevailing there. During 1975 two more probes were "soft-landed" (*Veneras 9* and *10*). From each a single picture was obtained. Both showed rocks everywhere within range of the camera. This proved also that the intensity of illumination on the surface of Venus was unexpectedly high. Wind speeds proved to be on the order of 10 miles per hour or less. This may seem very gentle, but it must be remembered that in an atmosphere as dense as that of Venus a gentle breeze would exert a considerable force.

It was left to the American probe, *Mariner 10,* during February 1974, to provide data relating to the cloud tops of Venus's atmosphere. Only one pass of Venus was made (*Mariner 10* was actually en route to Mercury). Nevertheless the results were outstanding, one might even say spectacular. They confirmed the fact that the rotation period of the upper atmosphere is only *four days* (compare this with the *planet's* rotation period of 243 days.) It is apparent, therefore, that the atmospheric structure of Venus is peculiar. Moreover the clouds are quite unlike those of Earth. There is evidence that they contain considerable amounts of sulphuric acid, a very highly corrosive material. If then these clouds produce rain, this rain must inevitably consist of droplets of dilute sulphuric acid—hardly the most pleasant form of precipitation. Whether or not such rain ever reaches the surface of the planet is still conjectural, but clearly it represents a considerable threat to probes which will have to pass through it and then perhaps be further subjected to it as they stand on the planet's surface.

About then it also became evident that the surface temperature must be in the region of 900°F. This is very much higher than all previous estimates. To this must be added the unpleasantness of a ground pressure some *ninety* times that of Earth at sea level. By now it must be all too evident that Venus, as presently constituted, represents no sanctuary for the peoples of a doomed or overcrowded Earth.

Why, it could very well be asked, should a planet of similar mass and dimensions be so very different from our own? The reason is not hard to find. Venus lies some thirty million miles closer to the Sun than does Earth. The surface has not therefore been able to cool down sufficiently. As a direct consequence, carbon dioxide has been expelled from carbonate-containing rocks. This has initiated the celebrated "greenhouse" effect whereby heat is retained. So, on Venus, began a highly vicious circle. The hotter the rocks became, the more carbon dioxide was released. The more carbon dioxide released the hotter became the atmosphere. If ever there was a vicious circle this was surely it.

Whether or not active vulcanism exists on Venus is still a matter of conjecture, since as yet we know very little of the planet's geology. This lack of essential knowledge concerning the planet's surface prevents our being able to paint a word picture of it. But of one fact we can most certainly be sure. It is a deadly world. We can safely visualize plains of tangled rock, the rock being perpetually eroded by tremendously powerful winds. Add to that the possibility of a rain of dilute sulphuric acid, and certainly a dense cloudy sky through which neither Sun nor stars can ever be seen. Venus is imprisoned in its own atmosphere and cut off, perhaps forever, from the universe beyond. As Venus is today we can only say "this is no place for us."

Mars: Outward from Earth we come next to the almost legendary planet Mars. Contrary to much popular belief, it is more distant from us than is Venus. It has, however, long been of greater interest because it is much more Earth-like than any of the other planets in the Solar System, despite the large disparity in dimensions (Mars has a diameter of about 4,000 miles, approximately half that of Earth).

At its brightest, Mars shines more brilliantly than any of the other planets except Venus. However it only achieves this distinction during favorably close oppositions which occur only every fifteen or seventeen years. At these times Mars is a magnificent object in our night skies. The memory of two such close oppositions remain very fresh in my mind—those of July 1939 and of August 1956. Perhaps the former was the more thrilling. At the age of sixteen it was still possible to dream of Martians.

With a diameter of 4,200 miles and a mass only one-tenth that of Earth, Mars is one of the smallest planets in the Solar System. Only Mercury and Pluto are smaller. The low mass means an escape velocity of only 3.1 miles per second (that of Earth is 7.0 miles per second). It is hardly very surprising, then, that the planet has been unable to retain an atmosphere comparable to our own. Nevertheless (and most surprisingly) the American Viking orbiters have shown evidence of once vast river systems. At least this is what they closely resemble. If they are the dried up drainage basins of river systems, then running water must have existed on Mars some tens of thousands of years ago. This, in turn, would indicate that at one time the Martian atmosphere must have been considerably denser than it is now. Today atmospheric pressure on Mars is so low that water cannot exist in the liquid state.

Mars takes the equivalent of 687 terrestrial days to orbit the Sun. This period thus represents the Martian year. In two respects the planet is very similar to Earth. The tilt of its axis is almost the same as ours. So also is its day at 24 hours 37 minutes. As a consequence of the tilt similarity, Martian seasons are of the same general type as our own, though of course, due to the longer Martian year, of greater duration. This is not to suggest however that spring, summer, autumn, and winter on Mars are in any way comparable to those seasons on Earth.

Before the first automated probes arrived in the immediate environs of Mars to scan its ochre and red surface, it was generally assumed that its topography was of a gentle undulating nature almost totally devoid of hills and valleys. It was also believed that the white pole caps were due to a thin layer of solidified carbon dioxide. Early results from *Mariner 4* in 1965 shattered these illusions. Mars was crater-scarred, a possibility that had never previously occurred to astronomers. It was also evident that the dark, shaded regions were *not* due to vegetation. Neither were they attributable to depressions on the planet's surface—a fact which finally laid to rest the "ancient dead sea bottoms" of Edgar Rice Burroughs' *Barsoom* (Mars). Indeed one of the most famous and conspicuous of these, the feature known as Syrtis Major, was in fact more like an elevated plateau.

Another unexpected result was the low pressure and low density of the Martian atmosphere. From this it was deduced that the major constituent of the Martian atmosphere was probably carbon dioxide, and not nitrogen as had hitherto been supposed. Ironically it was also revealed that the regions known as Hella and Argyre were, in fact, *depressed* features and not elevated areas as had so long been believed. Mars, it seemed, or at least our ideas concerning it, was apparently being rapidly turned upside down. Tantalizingly, other facts shone through. Mars, though a dead world and one far removed from the imaginings of Wells, Burroughs and a host of others, looked suspiciously line one which, aeons before, had been very different indeed. For example the Tharsis volcanoes appear to be associated with natural drainage systems. Indeed there are apparently many features that can hardly be regarded as other than dried-up river beds. Consequently the Martian climate must have been considerably less hostile in the past than it is now. The implications are intriguing.

The exploration of Mars was further advanced in 1976 with the landing of the two American Viking probes, each of which comprised an orbiting section and a lander. The first lander came down upon the plain known as Chryse and the second on another, known somewhat optimistically as Utopia. Both sites proved to be rock-strewn wastes, though there *is* evidence of the past action of running water. In both instances surface temperature of the planet proved very low, less than 22°F in fact. One of the principal objectives of both Vikings was to search for traces of any biological activity. This search has so far proved indecisive and the question as to whether microbic or other low-order life exists on Mars remains unanswered. But certainly no real live Martians strolled along to have a look at "the thing from Earth" which had just descended through their skies.

Under highly artificial conditions, small terrestrial colonies might be created on Mars. In this respect some of the well-known science fiction ideas might become reality—notably the transparent, domed, pressurized cities. But there would always be danger. In no way would Mars seem the answer to a dangerously over-populated Earth, and its thin atmosphere might be of little use against the dangerous radiation coming from solar flares. The value of Mars to the human race could lie in its mineral wealth, though the mining of metallic ores under such conditions would call for a high level of technology. And having obtained the mineral, there is the question of transport back to Earth and the relevant costs. Martian oil and Martian coal would hardly come cheap. The transport of Martian gold might make sense economically, though gold in itself is not a source of energy. In distress Mars could help us, but only a little.

The Asteroids: From our present viewpoint these can be disposed of quickly. They occupy that section of the Solar System between the orbits of Mars and Jupiter. They may be the remains of a large planet which somehow disintegrated. More likely they represent material that somehow never aggregated to form a planet. The four largest are Ceres, Pallas, Juno, and Vesta. Vesta is the brightest of the four, and is occasionally visible to the naked eye though its diameter is only half that of Ceres. There are thousands of these bodies and more are discovered every year. Some are so small they are not even spherical and are, in effect, just very large meteors. None of them could possibly represent a haven for even a minute proportion of the human race, though metals they contain might one day prove of inestimable value, assuming we have the technology to extract and transport them.

Jupiter: By the time we have reached Jupiter we are well into the cold, bleak outer regions of the Solar System where the Sun is beginning to look like precisely what it is—just a star. It still has a perceptible disc and gives out a measure of heat and light, but it is no longer the golden ball that floods our planet with light and warmth. Jupiter itself is the largest planet in the Solar System and possesses a magnificent system of at least 13 moons in orbit around it. Dr. Christine Sutton has described Jupiter as "lying at the heart of a miniature solar system." Her description could hardly be bettered. The four largest Jovian satellites are generally named the Galilean satellites, since Galileo discovered them with his early telescope. Each is very different. As Christine Sutton puts it "they include the youngest, the oldest, the smoothest, the most cratered and the most active surfaces seen so far in the Solar System."

Jupiter as a future abode for Earth peoples can be ruled out from the start. It is, in all probability, just an enormous spinning sphere of hydrogen and helium without a solid surface like Mercury, Venus, Earth, and Mars. All we see either by telescope from Earth or from probes is merely the upper atmosphere of the planet. The recent magnificent pictures sent back to us by NASA's *Voyager 1* and *Voyager 2* reveal all too clearly that the atmosphere is not only thick but extremely turbulent. On that day 5,000 million years hence when the Sun has swollen out to become a red giant star, consuming all the inner planets in the process, the Jupiter region is going to become a very much warmer place. But even if the astrophysicists had got their sums so wrong that the process started tomorrow the planet still could not be a haven to the human race. Jupiter, if it still exists in its present form in that distant era, will probably escape the engulfing insatiable menace of the Sun's outward expanding layers. What the roasting it will get will do to it is, at the moment, anyone's guess.

What could be of interest however are its moons, especially the four largest ones Ganymede, Callisto, Io, and Europa. The first three are larger than our own Moon and Europa is only marginally smaller. For the record, Ganymede is larger than Mercury while Callisto is virtually the same size. They are, in other words, really small *planets*. The other nine satellites are little better than large rocky lumps similar to the Asteroids. The largest of them, Amalthea (diameter 240 kilometers), is also the innermost. Io is the innermost of the large Galilean satellites (diameter 3,640 kilometers), and is presently the most volcanically active body yet found in the Solar system. (All future colonists to Io please note!) Io is almost the same size as our Moon. As *Voyager 1* flew by at least *eight* volcanoes were in full eruption. On the passing of *Voyager 2* seven were still going strong. Io has an atmosphere which contains sulphur, oxygen, and sodium. Our kind might appreciate the oxygen but not the sulphur and sodium. Volcanic materials are probably also responsible for the strange orange and red color of the satellite, sulphur being an element which can exist in different allotropic forms colored red, yellow, brown, etcetera. Active volcanicity, as might be expected, is also accompanied by evidence of past, extinct and dormant volcanic features. The surface of the satellite is, for example, dotted with large caldera ten to a hundred times larger than any known on Earth. A caldera is, in essence, a large depression brought about by the outpouring of lava from the magma chamber of a volcano. This empty chamber is eventually unable to support the rock above it and simply collapses. The photographs from *Voyager 2* also showed that volcanic vents are not simply confined to the equatorial regions, since some showed up at the south pole of Io.

Volcanic activity on this massive scale would indicate that the surface of Io is relatively young—probably about a million years old. This contrasts strongly with Earth where rocks at least one thousand million years old are known. Why should Io have such an active interior? One theory is that it is caused by the moon being pulled in one direction by the colossal mass of Jupiter and in the other by the relatively near moons, Europa and Ganymede. Despite all this vulcanicity, Io's surface is very cold (around -140°C). Certain "hot spots" were found to be 160°C warmer than their surroundings, which still only puts them around 20°C. This temperature is far too low for rocks to be in a molten state. The melting point of sulphur, for example, is in the region of 110°C. The heat is therefore well *inside* Io. Clearly this is no world for people bent either on colonization or escape from the Sun's wrath. For a "geologist's convention" it might be ideal, so long as they came prepared!

Europa is the next moon out from Jupiter and the smallest of the Galilean or major satellites. *Voyager 2* passed at a distance of only 206,000 kilometers. Pictures were good and surprise was great. Europa, according to Dr. Laurence Soderblom of the U.S. Geological Survey, "is as smooth as a billiard ball." This is not entirely true. Europa is undoubtedly very smooth but is, in fact, covered by a network of cracks. In the words of Dr. Christine Sutton "it resembles a cracked eggshell." The probability is that its surface is largely if not entirely water. Naturally this water is frozen, so that in fact Europa is covered by a layer of ice—possibly to the extent of 100 kilometers thick. No haven here then for the human race—not even for Eskimos.

Ganymede, largest of the Galilean satellites and therefore of all the Jovian moons is quite different from either Io or Europa. On this moon there are *two* distinct types of surface with abrupt changes between the two. There are dark, deeply-cratered regions and also younger, brighter terrain-containing ridges and grooves 5 to 15 kilometers wide and several hundred meters deep. If Europa is smooth, Ganymede is just the reverse. It looks inhospitable, and it almost certainly is.

The outermost Galilean satellite is Callisto. It is the least dense and so probably contains the most water. It is the most cratered body known and, in the words of Dr. Garry Hunt, "should keep the crater counters going for the next twenty years!" Since this points to meteoric impacts Callisto must be the oldest of the Galilean satellites, dating back to the period of heavy cratering four thousand million years ago. Here too is no place for the human race except perhaps a mere handful sustained under the most artificial conditions and totally dependent on Earth, a planet an average 400 million miles distant.

By now it must be becoming increasingly obvious that if for some reason Earth were ever to become uninhabitable, either by virtue of increased solar activity or human folly, there is little likelihood of finding salvation within our own system. Shortage of minerls or fuels might be alleviated but that is all. But round other stars swing other planets, some perhaps near replicas of Earth. There indeed could lie true salvation in some future age. The only problems are finding them and getting there.

However before we leave the Solar System for the unknowns of the interstellar leap let us take this present chapter to its logical conclusion—to the system's outermost members.

Saturn: Beyond the mighty Jupiter and its system of moons, moving at a mean distance of 886,000,000 miles from the Sun, lies another solar system in miniature: the lovely ringed Saturn and its family of ten moons. Like Jupiter its surface is gaseous, intensely cold, and there can be no question whatsoever of Saturn ever affording a refuge for the human race were it necessary to retreat somewhat from the more immediate environs of the Sun. It is a smaller planet than Jupiter with an equatorial diameter of 75,000 miles. Nevertheless it is still a colossus compared with Earth and its diameter of 8,000-odd miles.

There can be little doubt that the real glory of Saturn lies in its magnificent ring system. This feature may be due to the total disintegration of a satellite which ventured too close to the parent planet and suffered the inevitable consequences of such behavior. Alternatively the rings may simply represent debris which did not, for some reason, condense into a planet or moon. Whatever the reason, there is little doubt that the rings are composed of an infinite host of rock particles, some probably ice-coated. These rings lie in the plane of Saturn's equator as do the seven inner satellites of the planet.

The names of the satellites starting with the innermost and working outward are Janus, Mimas, Enceladus, Tethys, Dione, Rhea, Titan, Hyperion, Iapetus, and Phoebe. With the exception of Titan none of the satellites is very large. In diameter they range from the hundred miles of Phoebe to the eight hundred of Rhea. But Titan, as its name might indicate, is in a totally different class. It is the brightest, largest, and most massive of Saturn's satellites and the only one in the Solar System, thus far, proved to have an atmosphere. The diameter of Titan is at least 3,600 miles and though it orbits Saturn and must therefore, by definition, be termed a satellite, it really must, in some respects, be seen as a planet or pseudo-planet. The relatively small diameters of all the other satellites clearly rule them out as potential bases for terrestrial occupation and use, but Titan, on account of its size and the fact that it possesses an atmosphere of sorts, is clearly one that should be looked at a little more closely in this context.

During the year 1943-44 Gerard P. Kuiper photographed its spectrum using the 82-inch reflector of the McDonald Observatory in Texas. He found striking evidence of the presence of the gas methane in the Titan atmosphere and also a slight indication of ammonia. The satellite was originally discovered by the Dutch astronomer Huygens in 1655 and can be observed quite easily using a small telescope. Its diameter of 3,600 miles (the latest estimate) means that it is appreciably larger than the planet Mercury and is, in fact, not all that much smaller than Mars. It is certainly the largest satellite in the Solar System. Atmospheric pressure at its surface is about ten times as great as that at the surface of Mars. The Titan atmosphere is therefore relatively dense. It has been suggested that if we stood on the surface of Titan the sky would be blue instead of black, though this would be a deeper blue than that typical of the daytime skies of Earth. Because of its distance from it, the Sun would appear pale and wan, but from Saturn itself would emanate a strong, yellow light—reflected sunlight from Saturn rendered a more intense yellow by the effect of Saturn's atmosphere.

If all that has so far been said might make it appear that here at last is a place where some of Earth's population might exist, we must now display any such optimistic beliefs. There is no oxygen in the atmosphere and assuredly our kind cannot exist on a respiratory diet of methane and ammonia. Moreover the surface temperature on Titan is probably akin to that of Saturn itself, which is presently estimated at -240°F. Titan, it would appear, is rather a chilly place. Indeed it is probably only the intense cold that enables Titan to retain its atmosphere. Were the temperature at the surface raised by something like 100°F, the molecules of the gases would speed up to such an extent that the atmosphere would quickly evaporate off into space. It has even been suggested that Titan's atmosphere is continually being recycled, i.e., gas molecules escape from the low gravitational pull of the satellite but are prevented from being dissipated into space by the very strong gravitational pull of Saturn. They thus linger in orbit only to be recollected by Titan as it orbits the planet. This is an extremely interesting theory though it remains largely speculation for the present.

So far we are unable to say very much concerning the nature of Titan's surface. Drawings of Titan made by Dollfus using the 24-inch refractor of the Pic du Midi Observatory in the Pyrenees show light and dark areas. These are almost certainly indicative of permanent surface features. If vulvanism should happen to exist on the planet we might, it has been suggested, find "ice volcanoes" hurling fragments of frozen gases into the atmosphere and exuding a lava which was essentially a blend of liquid ammonia, methane and water.

Despite all the enormous difficulties and dangers it is considered in some quarters that a scientific base might one day be constructed on Titan. Such a base could be invaluable as a relay station for far-ranging probes. Contact with Earth could hopefully be maintained most of the time though not, of course, when Earth and Saturn lay on opposite sides of the Sun.

Such a base might be similar in design and form to those which one day may be constructed on the Moon. In the case of a Moon base there is a distinct heat problem during the two-week long lunar day, especially in equatorial regions. This problem at least, would not exist on Titan, for on Saturn's giant satellite the amount of light and heat received from the Sun only amounts to *one hundredth* of that received on Earth. This does not make for heat-waves! A Titan base would very probably be constructed underground, for the most part at least, but with a few observation domes or the like on the surface. Virtually everything would have to be shipped from Earth. Herein lies grave peril, for supplying material to such a base would be arduous and highly time consuming. Moreover if acute danger threatened, it would be totally impossible for a rescue operation to be carried out in time. Even a functional Mars base would still be very remote. In considering the ultimate vastness of interstellar space we must never lose sight of the fact that, though the Solar System constitutes only an infinitesimal part of this, it is still incredibly immense in respect to ourselves, our means of transport and of our communications.

In our brief updated tour of the Solar System we are getting steadily further from its central source, the Sun. The parts we have already reached might, in the case of a mildly flaring Sun, represent a haven from excessive heat and lethal radiation. Unfortunately we cannot suddenly become breathers of ammonia, hydrogen and methane—or even some blend of the three. The resultant rise in temperature would almost certainly dispel any such atmospheres possessed by large satellites anyway.

Uranus: Well beyond the orbit of Saturn, at a mean distance from the Sun of 1,783 million miles, swings the giant planet Uranus. The planet is only just visible to the naked eye depending on whether or not one knows where to look. Until 1781 the planet's existence was not even suspected. It was in that year that William Herschel discovered it, entirely by accident.

So far is the planet from the Sun it takes eight-four of our years to complete just *one* orbit around it—roughly the equivalent of a human lifetime. The period of its axial rotation is thought to be slightly less than twenty four hours. In 1977 Uranus stole a little of Saturn's thunder when it was discovered that it too was surrounded by a system of rings. These are much thinner than the rings of Saturn and are not visible from Earth. In terms of sheer magnificence they cannot be compared with those of Saturn.

Easily the most odd feature about Uranus is the curious tilt of its axis. That of Earth is 23½°, which is responsible for our well-known and well-defined seasons. Mars has a similar tilt and the other planets, though they all vary to some extent in this respect, are reasonably orthodox. But not so Uranus, the axial tilt of which amounts to a near fantastic 98°! This must lead to a very peculiar seasonal pattern. For twenty-one of our years much of the Uranian northern hemisphere will be in total darkness. For the following twenty-one this same darkness will descend on much of the southern hemisphere. However the terms "light" and "darkness" should not be construed by terrestrial standards. A Sun 1,783 million miles distant does not deliver a tremendous flood of light. The difference will probably be that between total darkness as opposed to a kind of grey twilight. Such distinctions are probably wholly academic since Uranus is quite likely to have no solid surface and, even if it has, its thick atmosphere is unlikely to admit much light.

Some idea of the actual constitution of Uranus may be gained from the fact that though Uranus has a volume *forty-seven* times that of the Earth its surface gravity is only *slightly* greater. In other words the density of the planet is low, so that it may well be composed wholly of gas. As a home for the human race it has no potential whatever. Science fiction writers of past years who gave us an inhabited Uranus and real live Uranians chose a most inappropriate venue for their tales.

Uranus possesses five satellites. At least it has at the moment of writing. With instrumented probes from Earth reaching further and further toward the outer frontiers of the Solar System another small and hitherto undetected one could be discovered ere these words are in print. Such is the pace of progress—and its repercussions.

The fifth and incidentally the innermost of these satellites, Miranda, was discovered by Gerard P. Kuiper in 1948. Proceeding outward from the planet we then come successively to Ariel, Umbriel, Titania, and Oberon. The two outermost are the largest with diameters of about 700 miles. Little is known as yet of their surface conditions but as habitats for human beings they are quite definitely out.

Neptune: By now we are 2,793 million miles from the Sun and have almost reached the known outer limits of the Solar System. Neptune was not discovered until 1846 but *not,* as in the case of Uranus, by chance. The discovery of Neptune was a triumph of mathematical astronomy with which the names of J. C. Adams and Urbain Leverrier will forever be associated.

Due to its immense distance from the Sun, Neptune is a very

difficult object for observation from Earth, and until an instrumented probe does a fly-by we are going to remain in a state of
considerable ignorance concerning it. Through a telescope surface details are very hard to discern, though there are apparently
bright and dark zones. The atmosphere appears to consist almost
exclusively of methane. The surface temperature is extremely
low, about -360°F. At this temperature any ammonia in the
atmosphere should have solidified. It has a distinctly bluish tinge
(appropriate to a planet named after the god of the sea), which can
probably be accounted for by the fact that methane absorbs red
and yellow light. Unlike Uranus it does not possess an extreme
axial tilt. In size it is similar but it is the more massive. There is,
of course, no question of human beings landing on this planet
either. In most internal respects it is probably very similar to
Uranus. What then of its satellites?

The planet Neptune possess two of these, named Triton and
Nereid respectively. Nereid is very small with a diameter of less
than 200 miles. Its main claim to fame is the remarkable eccentricity of its orbit; its distance from Neptune varies from 867,000
miles to an almost incredible six million miles. It is, consequently,
very much more remote from Neptune than is Triton. As an
observation base for the giant Neptune it would offer few, if any,
advantages.

Triton, the other satellite, comes into a totally different category. For a start it is one of the largest satellites in the Solar
System. Indeed some recent estimates have placed its diameter at
around 3,000 miles. It is certinly larger than the Moon and is
probably comparable in dimensions with Mercury. The satellite
orbits Neptune at less than 200,000 miles from the planet's upper
cloud layer. Its period of revolution is 5.75 days. It travels in a
retrograde or east to west direction. The orbit of Triton is also
steeply inclined (160°) which, in conjunction with Neptune's
axial inclination of 29°, means that the satellite passes high over
Neptune's northern hemisphere.

From Neptune and its two moons the Sun will appear comparatively small and its light feeble. None of the other planets will
appear brilliant in the skies of Triton and Nereid—not even its
"immediate" neighbor Uranus. In this cold and dark domain
there can be no refuge or alternative accommodation for the
human race.

Pluto: Since its discovery after long years of search by Clyde Tombaugh of Lowell Observatory near Flagstaff, Arizona, in 1930, Pluto has proved something of an enigma. Why this is so is discussed in the next chapter. It is, so far as is known at present, the outermost planet of the Solar System, though this statement calls for a little qualification. It so happens that Pluto possesses a most eccentric orbit which crosses that of Neptune (Figure 1). In

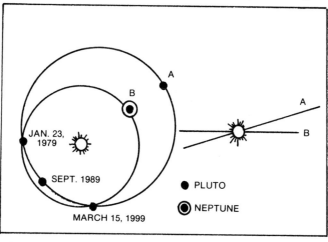

The orbit of Pluto. The eccentricity is relatively high, and at perihelion Pluto comes within the orbit of Neptune; perihelion is due in 1989. (A) Orbit of Pluto. (B) Orbit of Neptune.

Figure 1

this case there must be occasions when it is *closer* to the Sun than is Neptune. Pluto takes 248 years to make one orbit of the Sun but for 20 years it is *within* the orbit of Neptune, i.e., during its perihelion period. When first discovered in 1930, it lay mid-way between its furthest point from the Sun (aphelion, 39.4 astronomical units from the Sun) and perihelion (29.6 astronomical units). On the 23rd of January 1979, Pluto and Neptune were equidistant from the Sun at a distance of 30.3 astronomical units. Actual perihelion for Pluto is during September 1989. On March 15, 1999, the two planets will again be equidistant. On that date Pluto will regain the distinction of being the Solar System's outermost known planet.

Lest this crossing of orbits arouse fears (or hopes) of a collision between the two bodies, we must hasten to add that they are nowhere near each other at the "crossover" points of their orbits. Indeed, this is a state of affairs which can never be. The reason is a little involved. Astronomically speaking the two worlds are locked into orbit-orbit resonance. This really means that for every three revolutions of the Sun made by Neptune, Pluto makes two. The consequence of all this is that the two planets are closest together when Pluto lies at aphelion. Even then they are a comfortable eighteen astronomical units apart. We cannot therefore look forward to the celestial pyrotechnic display as Neptune tries to elbow Pluto out of its way. It goes without saying that Pluto, around five and a half light hours from the Sun, is no spot for occupation by the human race. A satellite named Charon was discovered in 1978. This enabled astronomers to fix Pluto's mass at approximately one quarter of the Moon's and its density at around 0.7g per cubic centimeters. This would appear to indicate a body composed entirely of materials found on Earth. On Earth however, these materials exist as liquids or gases. On Pluto, due to the very low temperatures, these materials must have solidified. It could be that Pluto has terrestrial features. It is certainly quite unlike its gas giant "neighbors" Jupiter, Saturn, Uranus, and Neptune. But even nitrogen and oxygen mixed in the proportions we call air on Earth would be no use to us—unless of course our respiratory systems somehow found themselves capable of handling solid air.

By and large the science fiction fraternity have left Pluto alone as the locale of "a great Plutonian civilization." I myself had a go at putting life on Pluto at the age of fourteen in what I hoped would be a science fiction epic. The creatures chosen as Plutonians were great slug-like creatures of limited intelligence. The story was never completed, and now I know better. (If, of course, huge slugs *are* ever found on Pluto I will immediately assume the mantle of prophet!)

What then has our brief tour of the Solar System shown us? The answer by now must be all too clear. Only Mars could offer anything to the human race by way of sanctuary and even there only for a relatively small proportion, under conditions of considerable adversity. The fundamental truth is that life can only exist in a comparatively small region around any star. That region is known as the ecosphere. Its extent and distance from the star depend entirely on the nature and size of that star. In the case of the Sun the ecosphere starts beyond Venus and almost includes Mars. Earth is more or less slap in the middle, which is highly fortuitous for us.

If therefore a mortal threat to Earth developed from a source external to the Solar System, say a "rogue" interstellar body (highly improbable) or by virtue of a "jay-walking" asteroid from within the Solar System (also improbable, but less highly so) then only Mars could serve as a refuge—and one that would be totally inadequate. If, on the other hand, the threat should originate in the sun as a consequence of extreme variability, extensive and prolonged solar flares or pulsating nova-like tendencies, only on the outer planets or rather some of their moons could there be safety. The big snag is that it would *not* be safety—only a choice in the way of dying. Moons like Titan and Triton and a planet such as Pluto have neither the atmospheres nor the other characteristics so utterly essential to our kind. And without them we die.

Man must eventually cross immensity to the stars. And even if no threat to Earth, Sun or Solar System compelled him he will still do it because that is just the type of creature man is. If for no other reason, he will go to the stars simply because they are there. But he can only do so if some other stars have planets.

③

$\mathcal{P}$lan εt 10

A S WE ALL know, the Solar System comprises one star and nine planets. Prior to William Herschel it was said to comprise one star and six planets. In those days the Solar System stopped at Saturn. It had been this way from time immemorial and there seemed no reason to expect it should ever be otherwise. But after Herschel's entirely unexpected discovery of Uranus the number of planets became seven. The brilliant mathematical work of J. C. Adams and Urbain Leverrier leading to the discovery of Neptune raised the planetary total to eight. Here it remained for several decades until the discovery of Pluto, in March 1930, by Clyde Tombaugh at the Lowell Observatory in Flagstaff, Arizona, brought the figure to the present total of nine. These successive "additions" to the Sun's family plus the peculiar circumstances relating to Pluto and its discovery have, rather inevitably, raised the question as to whether or not a tenth planet may lie out there in the darkness, as yet undetected. It must be obvious of course, that the Sun's dominion cannot extend indefinitely into space. Nevertheless the possibility of at least one more planet lying out and beyond the orbit of Pluto, perhaps even a major one, is a distinct possibility. Tracking it down, however, could prove a far from easy task.

In the circumstances it seems not unreasonable to ponder the matter of a probable "Planet 10" before proceeding to our main theme, the planets of other stars. In a sense this probably represents a good bridge between our system and others.

In the circumstances a few words concerning the enigma of Pluto may be appropriate. Percival Lowell believed that the planet lying out beyond the orbit of Neptune and responsible for the perturbations in the orbits of Uranus and Neptune should have a diameter of 25,500 km (16,000 miles). This figure was based on the assumption that the unknown planet would have a similar density and composition to that of Neptune. He also believed that the albedo of the planet (its light-reflecting capacity) would be of a similar order. Unfortunately when at last in 1930 the planet was discovered (several years after Lowell's death) it was found to be much *fainter* than Neptune. Though this fact raised certain doubts in the minds of some astronomers, the planet subsequently named Pluto (after the god of the underworld and of darkness) appeared to fulfill the predictions made both by Lowell and by Pickering. In view of the crude and ofttimes uncertain data with which these astronomers were compelled to work, too much should not be made of this fact. In 1950, five decades after its discovery, Kuiper and Humason, using the 200-inch telescope of the Mount Palomar Observatory in California, were able to place a figure on the diameter of Pluto— 5,800 km. (3,600 miles). Thus, if the figure was correct, Pluto was *smaller* than the Moon! This contrasted very oddly indeed with the predicted diameter of 25,500 km. (16,000 miles) on which the calculations leading to the discovery of the planet had been based.

In 1977 a much more accurate estimate of the planet's diameter was made by Cruikshank. As a result the mystery deepened even further. Pluto, it now appeared, had a diameter of only 2,700 km. (1,700 miles). In 1978 a satellite to Pluto was detected by Christy and given the name Charon. This enabled an upper limit to be placed on Pluto's mass. This worked out at only 1/383 that of Earth. The density of Pluto could thus only be 1.0 or even less, a fact which led to its being described as "just a very large snowball." This has by no means proved the last word in the sad story of Pluto's downgrading, for Brian Marsden of the Smithsonian Observatory has postulated that Pluto is really nothing more nor less than a very large planetoid. Marsden bases his case on the strong parallels in orbital behavior between Pluto and Chiron, the slow-moving planetoid discovered between Uranus and Neptune by Kowal in 1978. In short then, the mass of Pluto is so low that it could not possibly be responsible for the perturbations in the orbits of Uranus and Neptune on which Lowell and Pickering's work was based. The mass allowed for by Lowell and Pickering was six times that of Earth. In the light of recent discoveries concerning Pluto it is apparent that the figure is 0.002—about *three thousand times less.*

The implication therefore can only be that, in March 1930, Tombaugh simply got lucky. Such things happen to astronomers as well as to other mortals and certainly the fact should not be regarded as adverse criticism of Tombaugh, whose initiative, perseverance and skills were of a very high order. After all, as it later transpired, images of the planet had appeared on photographic plates made by Lowell and others years before—images which had escaped their notice. Perhaps Tombaugh was searching in the right area of the sky but just found, by a curious mischance, the wrong object.

Thus it would appear that the real source of the perturbing influence on Uranus and Neptune remains to be discovered. Mathematically that source ought to be represented by a planet of considerable mass (several times that of Earth)—in other words Planet 10. Where then do we presently stand in our reasoning and in our search for this dark, mysterious and, as yet, hypothetical world?

In 1972, Carpenter and Brady worked out the ephemerides (position and other numerical parameters of a celestial body at a given time) of Planet 10, basing these on the orbital motion of Halley's Comet. (Their actual aim was to forecast accurately the perihelion date of that comet on its return in 1985-86.) Brady and Carpenter's computations required the hypothetical planet to orbit the Sun at a distance of 63.5 astronomical units (one astronomical unit or A.U. is equivalent to the distance between Earth and Sun of 91 million miles). It should also, they reckoned, have an orbital period or "year" equal to 464 terrestrial years. The mass of the planet was estimated as some three times that of Saturn or about 0.9 times that of Jupiter. Unfortunately, for mathematical reasons (which we shall certainly not go into here), this would also have called for retrograde or reversed motion and an orbital inclination of 120° to the plane of the ecliptic. This naturally aroused a considerable amount of scepticism. In 1973 Kiang showed that the existence of a trans-Plutonian planet did not really enter into the subject of forecasting the perihelion date of Halley's Comet, this being entirely due to interaction of forces existing between the Sun, Jupiter, and the comet itself.

The finding of Pluto was, likely as not, an accident and a highly unfortunate one at that, since in all likelihood it was a red herring which has prejudiced and delayed search for the real perturbing planet. In this respect however the air now seems to be clearing and it is possible that a search for Planet 10 may eventually get under way in earnest. The fact that Pluto may only be a large asteroid and that an asteroid has recently been found between the orbits of Uranus and Neptune might conceivably indicate that the well-known asteroid belt lying between the orbits of Mars and Jupiter may not be the only such belt within the Solar System. Is it possible that asteroid belts of lower density lie between the orbits of Uranus and Neptune and perhaps between those of Neptune and Planet 10 as well? At such a range the individual members would be very difficult to discern from Earth.

It is an accepted fact that planetary orbits tend to coincide with the aphelions of comets (i.e., the points at which comets are furthest from the Sun). For example Halley's Comet and about six others reach out to the orbit of Saturn before turning back and heading sunward. Jupiter's orbit, for its part, marks the outer or aphelion limit for another fifty comets. This is not due to coincidence but is simply a practical manifestation of the immense gravitational attraction of the planets in question. Other comets reach out beyond Pluto before turning back toward the Sun. It is clear by now, however, that Pluto's gravitational attraction is simply too slight to achieve this. The inference must be that an extra-Plutonion planet exists which is having this effect on comets, the aphelions of which lie beyond the orbit of Pluto. The distance is reckoned to be in the vicinity of 75 A.U.s which represents the next niche in Bode's famous Law governing the distance of each planet from the Sun. Present belief is that Planet 10 is not a gas giant after the fashion of Jupiter, Saturn, Uranus,

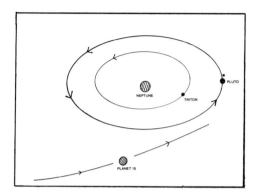

BEFORE

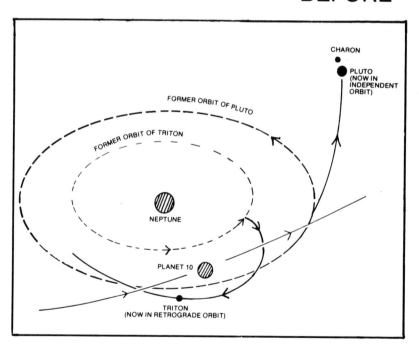

AFTER

Figure 2

and Neptune but a solid planet considerably larger than Earth or Venus. Assuming a density of 2.0 its diameter would be in the region of 25,000 km. (15,500 miles).

Recently some interesting speculations have been made regarding the early history of Planet 10, though it must be stressed that these *are* only speculations. Robert Harrington and Thomas van Flandern, both of Lowell Observatory, Flagstaff, Arizona, suggest that the Solar System's tenth member (having 3–4 Earth masses) passed very close to Neptune thereby wrenching Pluto (formerly a satellite of Neptune) free. The tidal forces which resulted were responsible for creating Charon (the newly discovered satellite of Pluto). They also imparted to Triton (the satellite of Neptune) its present retrograde orbit. The overall effect of this dramatic and disastrous encounter was to remove Planet 10 to a new orbit even further out, probably 50 to 100 A.U.s from the Sun, thereby reflecting so little of the Sun's light as to be extremely faint (Figure 2). Alternatively Pluto might have been a satellite of Planet 10, gaining its independence in the encounter between Neptune and Planet 10.

If the speculations of Harrington and van Flandern have any validity the distance quoted earlier of 75 A.U.s might have to be revised. According to Bode's Law the distance of Planet 10 from the Sun ought to be 77 A.U.s, so that the two are in very close accord. However it is accepted that the orderly relationship represented by Bode's Law has little validity beyond 30 A.U.s. This apart, it would certainly be lost after the kind of celestial billiards game envisaged by Harrington and van Flandern.

The degree of brightness likely to be displayed by this hypothetical planet is very much a matter of conjecture. It is also a matter of paramount importance when it comes to locating the planet. Brightness as such depends on three factors: (a) distance of planet from the Sun, (b) diameter of the planet, and (c) the nature of its surface. If something along the lines suggested by Harrington and van Flandern did take place, the orbit of Planet 10 could be very eccentric, intersecting that of Neptune just as Pluto does. The overall picture is of necessity hazy, but might conceivably be of a planet having a mass five to six times that of Earth, an orbital period or "year" of some 500 terrestrial years and lying at a *mean* distance of 63 A.U.s from the Sun. At perihelion (30 A.U.s) the planet would be about a hundred times brighter than when at aphelion (93 A.U.s).

The diameter of a planet depends on its density, and so if the density of Planet 10 approximates to that of Jupiter it would have a diameter of the order of 24,000 kms. (15,000 miles). This is roughly equal to half the diameter of Neptune. It would also possess about 25 percent of Neptune's reflecting surface. Should the albedo be about the same for both planets, Planet 10 would be four times less bright. The maximum magnitude or brightness of Neptune is +7.8 at perihelion, therefore the magnitude of the unknown planet would be about +9.3 (+14.2 at aphelion). For those unfamiliar with values for brightness (or magnitude as it is termed by astronomers) it should be stated that the scale starts at zero and runs in two directions: increasingly negative values are indicative of increasing brightness, increasingly positive values are indicative of decreasing brightness.

When Pluto was discovered in March 1930, its magnitude was only +14.5. Why then, if Planet 10 exists and has the magnitude attributed to it, has it not already been found? The answer could be that its albedo (i.e., its capacity to reflect the Sun's light) is not as high as that of either Neptune or Pluto. In other words its surface composition must be very different from that of Neptune or Pluto. Alternatively the planet could be denser than Neptune. It would therefore be smaller and provide less light-reflecting surface.

It is interesting to speculate in what region of the sky the planet, if it exists, might be found. Present thinking, having due regard to such facts as are available, indicates that the planet might presently lie in the constellation of Aquarius, a region of the sky well placed for observation during the evenings of late summer and early autumn. We must stress however that only telescopes of fairly considerable aperture and power are suited to such a search. Photography must also be employed. It is, in short, a job for a properly equipped observatory. Unfortunately at the present time most observatories are engaged largely in stellar and cosmological work, so that if a Planet 10 really exists it may be able to escape the gaze of mankind for some time yet.

It must be strongly emphasized, however, that there is no certainty that such a planet exists. Nevertheless the fact that the perturbing influence exerted on Uranus and Neptune cannot adequately be explained by a planet as small as Pluto, is certainly a strong point in favor of the existence of the planet. Pluto undoubtedly represents an anomaly. Though it is now a planet in its own right, its pedigree is, at best, highly suspect. As we have already seen it could be a former moon of Neptune elevated unexpectedly and violently to the undeserved status of planethood. Its eccentric orbit which intersects with that of Neptune is merely further evidence that Pluto has achieved a rank in the Solar System to which it has probably no real title. Indeed, the more we learn of the outer reaches of the Solar System the greater becomes the suspicion that some rather queer happenings have taken place there. Pluto is too small, its orbit is too eccentric; inexplicable perturbations affect the motions of Uranus and Neptune. Now we find that Pluto has a moon and that Uranus has also a system of rings, though of a less grandiose nature than those of Saturn. The dispatch of unmanned probes and later manned vehicles to these regions may lead to many more surprises. They may also lead to the belated discovery of Planet 10. Such a planet is almost certainly no use to us as an alternative abode, and if our Sun went wild it would probably still be too close to it to represent a safe haven for humanity.

$$\textcircled{4}$$

$\mathcal{In}$ $\mathfrak{the}$ $\mathcal{Beginning}$

N OW LET US investigate the manner in which planetary systems may be formed. To go into this aspect in any detail it is
clearly essential to have as much data relating to planetary
systems as possible. And this, at once, is where the difficulties
arise, and for the most obvious of reasons. At present we have
simply no direct, intimate knowledge of stellar planetary systems. Even with our largest telescopes we cannot hope to detect
planets as large as Jupiter orbiting any of the nearest stars. In the
following chapter we will be looking at what is by now virtually a
existence of stellar planetary systems can be postulated, and in
chapter 6 we will be looking at what is by now virtually a
confirmed *other* solar system. But we will also see just how very
difficult it is to come up with precise details about such systems.
How then can we ponder possible mechanisms to account for
their formation?

In a sense the answer is really very simple: We sit in the middle
of one. We can examine it in detail, at leisure and now, thanks to
the advent of space exploration, we have at our disposal a wealth
of information long denied us. Some readers may feel this is all too
convenient and neat, and raise that other awkward question—is
the Solar System typical, a prototype? Might it not be unique?
This is a perfectly fair question and one that we will be dealing

with later. For now let's just say that there *are* reasons, fairly good and sound ones, for believing that our Sun and its retinue of planets, moons, asteroids, and comets are *not* unique. More than half of the stars we see are binaries or multiple, i.e., pairs or groups of physically related suns. We know also of stars with companions only 0.016 as massive as the Sun, or about seventeen times the mass of Jupiter. It seems reasonable to suppose therefore that many stars are accompanied by bodies of planetary dimensions. All this we will be examining. For the present let us say again that our solar system is almost certainly not unique. This is not meant to suggest that all solar systems will be alike. Ours will certainly not be a prototype in the accepted sense of the word. In chapter 6 we will be presenting a case for a planetary system around Barnard's Star which lies about six light years distant. It is a reasonable assumption that this system of worlds does not contain nine major planets as does ours. And even if it did they would almost certainly be spaced differently and be of differing dimensions. It *is* nevertheless a solar system. Our own system we can regard as a prototype only in the sense that planets and other lesser bodies orbit a star. This must be our definition of prototype in the present context.

The very systematic arrangement and motion of the planets and asteroids constitute impressive evidence that they represent a true system. It seems virtually inconceivable that such an arrangement could possibly have been the consequence of chance encounters between all these bodies. Let us look at the question another way. The nearest star to the Sun, Alpha Centauri, is about 60,000 times as remote as Pluto, outermost planet of the Solar System. This body could hardly have played a part in the proceedings. The members of that assemblage we call the Solar System must then have had a *common* origin. Over the years there have been several theories to account for that origin but all fall into one or other of two easily defined categories:

a. origin by a process of orderly evolution, and
b. origin by virtue of catastrophe, i.e., some unexpected and unpredictable large scale event.

To gain a clear insight into the whole question it is necessary to review the history of the matter. A classic example of the first type is that due to the celebrated French mathematician and astronomer Pierre Simon Laplace who, in 1796, published his renowned nebular hypothesis. Laplace postulated a vast disc-shaped mass of cold gas in slow rotation which originally extended beyond the bounds of the Solar System as we know it today. As this cloud of gas (nebula) contracted under the mutual gravitation of its parts its rate of rotation, in order to conserve angular momentum, had of necessity to increase. Angular momentum of a body rotating around a center can be expressed as the combined product of its mass, velocity, and radius. In mathematical notation this is generally denoted by the expression

$$L = m v r$$

where L is the angular momentum, m is mass, v velocity and r radius. It can be shown that the angular momentum of a closed system cannot change. By referring to the equation above we can see that if r (the radius) *de*creases, v (the velocity) *must* increase proportionatly. Mass (m) of course, remains constant.

Eventually, according to Laplace, velocity of rotation (v) increased to such an extent that centrifugal force at the periphery or rim regions of the disc-shaped nebula *exceeded* that due to gravitation, causing a ring of material to separate from the main body. As contraction continued apace, i.e., as r, the radius, continued to diminish, other rings of material broke off at successively smaller distances from the center.

Laplace postulated further that these rings were not of uniform width throughout. The thickest portion gradually drew material toward and into it. This in the fullness of time condensed into a planet. In turn, and by an analogous process, satellites were formed as the planets contracted. Comets, meteors, and other bodies could be regarded as debris, mere waste material. The Sun, of course, represented the remains of the primordial nebula.

Laplace's nebular hypothesis must certainly rank as one of the greatest scientific theories of all time. It endured for the better part of a hundred years and even its demise, when it came, was a fairly protracted business. That demise centered on the fact that the planets, though possessing barely 1 percent of the Solar System's mass, had somehow contrived to acquire 98 percent of its angular momentum! The odd feature here is not so much the fact that the planets have this much momentum, for in truth it is just enough to keep them in orbit in their present positions. The anomaly lies in the fact that the Sun so *lacks* it.

It is well nigh impossible for angular momentum to be so ill-distributed. If the Solar System condensed from a gas and dust cloud with sufficient spin to account for the angular momentum possessed by the planets, the Sun should really be spinning about a hundred times as quickly as it is. Is it possible that it *did,* in fact, have this degree of momentum initially and somehow shed most of it?

Here was an interesting problem. One astronomer who suggested a possible solution was Hannes Alfvén of Sweden. Magnetic interaction, he proposed, between the Sun and a cloud of electrically charged particles succeeded, over a long period of time, in producing the necessary braking effort. Fred Hoyle of Britain came up with a very similar proposal in 1961 by suggesting that, as the Sun in its initial formative phase contracted within what is now the orbit of the innermost planet Mercury, it shed gas. Within this gas, large chemical aggregates formed which were held in the magnetic web generated by a very hot, fast spinning Sun. As these aggregates, rather like large boulders, spread outward they remained held and so carried with them a high proportion of the Sun's angular momentum.

In a sense however, we are getting ahead of our story, for these suggestions to explain the enormous disparity between the angular momentums possessed by Sun and planets respectively came much later. Laplace's theory began to founder over this point toward the turn of the century, whereas the proposals put forward by Alfvén and Hoyle belong to a much later era.

What happened next was, now that it can be seen in retrospect, a sort of unfortunate natural consequence. Alfvén and Hoyle sought an answer in an orderly continuity of events. Those who pondered the matter at the dawn of the twentieth century did not. The reason for the doubts concerning Laplace's hypothesis, as already stated, was the fact that the planets possessed so much of the Solar System's angular momentum and the Sun so little. Concentration was centered more on the amount possessed by the planets than on the small amount retained in the Sun. At any rate, there was a most marked imbalance in the system and the logical conclusion could only be that something had put it there. Here indeed was the fork in the road, and unfortunately the wrong path was chosen. No *normal* physical process, it was decided, could have been responsible for it. Laplace's hypothesis had looked favorable but for a single anachronism. Somehow that anachronism had to be explained. The pity is that, in doing so, the theorists of the time were prepared to abandon Laplace's hypothesis entirely.

It was now argued that the necessary energy must have been forcibly injected into the system by some outside agency and that agency could only have been another star. Somehow another star had supplied the energy and the only way this could have been achieved must either have been by collision with the Sun, or perhaps more conveniently, by virtue of a *near* collision. Thus was born the celebrated "other star" theory which was to dominate thinking in this field for about four decades. It is perhaps fair to state that the idea was not wholly new, for as early as 1745 a French naturalist, one George-Louis Leclerc, Comte de Buffon, had proposed that a comet had struck the Sun aeons ago and knocked off the lumps which eventually became the planets. There seems to have been considerable confusion in the noble count's head regarding the disparity in dimensions between a comet and a star. However it was still only 1745, de Buffon was a naturalist, not an astronomer and so, in the circumstances, he can be forgiven.

It is interesting to examine the sequence of events postulated in this theory, which, incidentally in all its variations, is due largely to four men, T. C. Chamberlin and F. R. Moulton in the United States and Sir James Jeans and Sir Harold Jeffreys in England. As the other star approached great tides were raised on the Sun—and, we would imagine, on the intruder star also. As the latter drew steadily closer the tides on the Sun were transformed into a great bulge which in turn led to an ejection of matter from the Sun in the form of a long cigar-shaped filament of matter. Since the outer extremity of this (it was argued) followed the passing star into space it must have moved in a *curved* direction around the Sun. Some of the ejected material fell back into the Sun. In this way the latter was induced to rotate in the same direction as that of the filament around it. The filament then gradually disintegrated into separate parts which in the fullness of time condensed into the planets. Once again, as in Laplace's theory, comets, asteroids, and meteors represented debris and leftover material.

At first glance this seems a reasonably plausible theory, and indeed it endured for many years. The first serious cracks appeared only when it began to be subjected to a strict mathematical analysis. Calculations indicated with increasing emphasis that the ejected filament left conveniently orbiting the Sun would, almost certainly, have *followed* the intruder star as it passed on into space. From the point of view of the theory's protagonists this was highly unfortunate, since in so many ways it seemed to fit the circumstances so well. For example, the cigar-shaped filament was at its thickest where Jupiter, greatest planet in the Solar System, now lies. Going spaceward it then tailed off nicely giving a good "fit" for Saturn, Uranus, and Neptune. At the "sunward" end the fit was slightly less than perfect, though the innermost planets, Mercury, Venus, and Earth, were accommodated very comfortably. Mars was too small however, while between it and Jupiter there occurred a large and seemingly inexplicable gap. This, it was felt, might once have been occupied by a larger, "good fit" planet which for some unaccountable reason broke up to become asteroids—or so the hopefuls postulated.

As the thirties drew toward their close, more and more objections were being raised against the "other star" theory. In 1939, for example, Lyman Spitzer, Jr., of Princeton University drew attention to the fact that hot gas drawn from the Sun would *not* condense into planets. Rather, it would spread out around the Sun as a tenuous cloud. And again it was being pointed out by other scientists that the filament of matter supposedly pulled from out the Sun would be much more likely to have followed the other star into space.

To counter the latter objection the theory was extensively modified. Now we were asked to assume that the Sun had once been a component of a binary or double star system (these are very common in space and we will be considering them in some detail when we deal with the possibility of planets in such systems). It was this companion of our Sun which, it was argued, had been struck by the intruder star. The colliding stars then both went off in different directions, leaving the central portions of the resulting filament in the vicinity of the Sun. This idea, it may be of interest to add, was originated by Henry N. Russell of Princeton University, an astronomer famous for his work in the classification of stars.

This represented a highly speculative modification of an already speculative and fairly colorful hypothesis. Moreover, though it countered one objection, it did not resolve that raised by Spitzer for, in this instance too, highly heated material extracted from the thermal inferno of a star's interior would only have dispersed itself into a tremendous halo around the Sun.

"Other star" theories have also one inherent and very fundamental weakness. It is the undeniable fact that stars are separated from one another by distances so immense that the chances of stellar collisions are virtually nil. Indeed the possibilities have been likened to dropping three tennis balls into the Pacific Ocean at widely separated points and expecting a collision to ensue between any two of them. Theoretically such a collision *could* happen. In practice we know very well that it will not.

Belated recognition of this apparently obvious fact led to a further "catastrophe" theory. By its terms a filament of matter is said to have originated when the Sun's companion star became a supernova or exploding star. This is a nuclear explosion on a scale so grand that there are virtually no superlatives to describe it. At times the event renders a star brighter than an entire galaxy. Such cataclysms are, however, very rare indeed and only about six have been observed in our own galaxy during the past 2,000 years. Occasionally one can be spotted in another galaxy. Since all other galaxies are more than two million light years distant (and generally much further) the really awesome character of a supernova can be gauged.

The "supernova" theory has perhaps a greater degree of probability than that of the "passing star." In terms of chance this is only relative as the above statistics show. Moreover surely an exploding star would throw out its material in *all* directions and not just in such a direction that it could be conveniently trapped and made into a planetary retinue by the Sun. But then perhaps it was only the material sent hurtling toward the Sun which was ensnared? But if so, how was it fashioned into a suitable filament? Question follows question as credibility, already strained, begins to wane.

In the early forties C. F. von Weizsäcker and Otto J. Schmidt reverted to the idea that in some way or other the Solar System had been created from a cloud of dust and gas. Von Weizsäcker suggested that vortices were formed systemmetrically in a primordial cloud that was already girdling the Sun. In time these vortices condensed into the planets. Schmidt claimed, however, that the planets were in fact "captured" by the Sun from clouds of dust and gas known to exist in interstellar space. In Holland, H. P. Berlage revived Laplace's earlier theory of two hundred years before, claiming that rings formed spontaneously in the original nebula. However, all these theories were criticized on the grounds that particles would not accrete in this way to begin the formation of planets.

The "cloud" theory that presently appears to satisfy all the necessary conditions, so far as they are known, was proposed back in the fifties by the American astronomer Gerard P. Kuiper, who claimed that the vortices suggested by von Weizsäcker were too uniform and that the more likely cause was turbulence in various parts of the cloud where material became sufficiently dense to accrete despite the gravitational influence of the Sun. This effect was enhanced, indeed made possible, by random, swirling movements of gas within the cloud. This movement, becoming progressively more rotational, had the effect of flattening and condensing the cloud into a disc-shaped mass of matter, in the middle of which were one or more large concentrations of material with smaller ones streaming around them. Subsequent stage was the formation of a proto-sun at the center. This would probably indicate that, so far as the Solar System is concerned, there was only one concentration of matter at the center, hence one star which we call the Sun. Where two or more such concentrations are concerned, binary or multiple star systems respectively are probably the result.

By this time the Sun or central star had condensed sufficiently for nuclear reactions to begin. It therefore commenced to glow and the heat and other forms of radiation emanated as a result, began to disperse the gases still surrounding the protoplanets. As the radiant power of the new-born star grew, gas shells around the innermost planets were entirely dispersed leaving only bare cores. The larger, outer planets were less affected in this way. Ultimately, increasing radiation from the maturing Sun dispersed and removed the last of the free gases still whirling around the system. The end result was a retinue of planets, those near the Sun being small and solid (i.e., Mercury, Venus, Earth, Mars) and those further out much larger and gaseous (i.e., Jupiter, Saturn, Uranus, and Neptune). Pluto, as we have seen, is something of an anomaly.

It is, at best, extremely doubtful whether any cosmologist would be prepared to endorse this theory as perfect. On the whole however, it gives the impression of representing a *likely* sequence of events. These events may not be precisely as we have described them. Perhaps part of the sequence is missing. But the concept, taken as a broad whole, does seem eminently more rational than one which demands, against tremendous mathematical odds, the close approach of one star to another, a stellar collision or one of a binary pair becoming a nova. Above all, the concept means that planetary systems can be regarded as the norm, a total reversal of what acceptance of "catastrophe" theories imply.

Kuiper's theory implies that star and planets are created more or less at the same time, give or take a millenium or two. This is somewhat at variance with the belief that a star is born or at least a proto-star comes into being, from which subsequently belts of matter are flung off to condense eventually into planets, thereby robbing the star of its former extremely rapid rate of rotation. This is something we will be looking at more closely in the next chapter, but some readers will probably have noted the marked similarity in this to a facet of Laplace's original nebular hypothesis. Whether or not this concept can be reconciled in any way with Kuiper's theory remains to be seen.

We know of the existence of one stellar planetary system with total conviction—our own. Cosmological arguments aside, it would seem reasonable to assume that what has happened to one star, and a very average and undistinguished one at that, must happen to a proportion of like and near-like stars in this galaxy and in others. The idea of our Solar System being unique in the galaxy, even in the universe as a whole, seems both irrational and absurd. What has happened here must have happened and be happening at a host of other points in space. This is the story we now take up in subsequent chapters.

$$\left(5 \right)$$

Piercing the Veil

TO REMARK THAT the detection of extra-solar planets is extremely difficult now is merely stating the obvious. Very briefly, we will again outline the complexities and their implications. The primary factor, the fundamental parameter, is of course, distance which, even in the case of the nearer stars, is of a kind that simply defeats the imagination. The nearest star to our own Sun is the triple system of Alpha Centauri 4.3 light years distant. Translated into miles and the figure rounded off this represents a great abyss of 25 million, million miles (or 25×10^{12} miles using the convenient notation of the mathematician). It is no good trying to visualize such a distance. Even one light year (about 6 million, million miles) is impossible to envisage. Indeed we cannot even comprehend a single million miles. Could we produce a space vehicle with a velocity of one million miles per hour (and such a remarkable achievement is not even remotely in sight) a one-way journey to the Alpha Centauri system would occupy something of the order of three thousand years which, in view of the relative brevity of the human life span, is rather on the long side! Such a vehicle would wind up merely as a travelling inter-stellar morgue. We, or presumably what was left of us, might indeed make Alpha Centauri but we certainly would not be able to enjoy any of its delights—assuming it has any.

When in our opening chapter we took a brief but close look at our own Solar System, it appeared that Earth and its eight sister worlds are well separated from the Sun, some more than others. Indeed so far out is Pluto that its light (which is merely reflected sunlight) takes about five and a half hours to reach us. That is quite a thought—five and a half hours for the light from the Sun to reach Pluto and a further five and a half hours for it to return. That is a round trip time of over eleven hours for light, even in the confines of *our* relatively small corner of space.

Surely then it might be argued that alien astronomers on the planets of "near by" stars should at least be able to detect the large outer planets, Jupiter, Saturn, Uranus, and Neptune by optical means, i.e., giant telescopes on the Mount Palomar pattern. If we happen to think along such lines we are merely deluding ourselves, to be fooled by the fact of our existence within the Solar System—a kind of cosmic insularity. The truth is that these four great outer planets of the Sun, viewed at a distance of only 4.3 light years (the distance separating us from Alpha Centauri), would be completely undetectable by optical means. Compared to distances of an interstellar order the distances between our Sun and its planets are minute. And though the four planets we have mentioned are undoubtedly vast, they are vast only in comparison to our relatively small Earth. Compared to the Sun they are very small indeed (just as our Sun is a pygmy of a star compared to the super red giants Betelgeuse and Antares). At a distance of 25×10^{12} miles (4.3 light years) these planets would not just appear small, they would *not* appear at all! To all intents and purposes our Sun would give the impression of being a solitary, yellow star. Even if an optical telescope of sufficient power to detect the planets of the Solar System could be constructed, they would still remain undetected since they would be lost in the glare emanating from the Sun. Thus any idea of observing extra-Solar planets from the surface of the Earth is a total non-starter and it is certain that neither time, money, nor resources should be expended on so futile an exercise.

The landing of men on the Moon by the United States again raises the possibility that, in time, a large optical telescope erected on the Moon's far side, eternally free from "Earth shine" and not disadvantaged by a deep, turbulent atmospheric ocean to penetrate, might just be able to detect large planets orbiting some of the nearer stars. This still seems more like pious hope than a serious attempt to come to grips with reality. The absence of a thick, turbulent atmosphere would certainly work in favor of such a scheme but the relative smallness of even giant planets at such colossal distances, coupled with glare from the parent star's close proximity, still applies and would be unlikely to be overcome despite the much more favorable circumstances.

If then we are going to detect the presence of extra-solar planets it will have to be by methods other than optical. And *detection* really is mandatory. Theorizing the possible existence of such planets is all very well, but it cannot and does not supply absolute *proof* of their existence. The theory referred to is two-fold in nature:

1. Some stars have a much slower rate of spin than others. This could be attributed to the birth of planets and a consequent loss of much angular momentum. Stars to which this applies are older and generally of a more appropriate type. Young and very hot stars are fast rotators that have probably not so far spawned planets (or may never be going to!). Stars with slow rates of rotation (and our Sun comes into this category) can thus be regarded as *possible* centers of planetary systems.

2. In cases where a star's proper motion deviates from a straight path, the deviation may be attributable to the perturbing influence of dark, unseen companions, i.e., large massive planets. The stars so affected are slow rotators like the Sun. Detecting this "wobble" in the true path of a star involves the application of astrometry, that branch of astronomy concerned with the precise determination of the *position* of stars. Superficially it might seem that this involves a fairly straightforward technique and a fairly satisfactory one despite the fact that we do not actually "see" the planet(s) as such. Simply examine all slow-rotating stars for "wobble" and all is well! Alas the matter is not that easy. Some idea of the magnitude of the task may be gained if we consider a practical example.

Let us assume that around the star system nearest to our own

(Alpha Centauri at 4.3 light years) there are planets, and on one of these planets there are astronomers who, having reasoned as we have done, along the lines of slow stellar rotation and proper motion deviation, decide to examine *our* Sun for planetary presence. Now we would not be so foolishly optimistic as to assume that the planets Mercury, Venus, Earth, or Mars would impart sufficient deviation to our Sun's predestined path through space for this to be detectable at a distance of 4.3 light years (25×10^{12} miles). Surely though, the giant planets Jupiter, Saturn, Uranus, and Neptune, so immense compared to the tiny inner planets, should reveal their presence in this manner. The verdict in this instance may prove surprising, for it has been calculated that detecting the wobble in the Sun's motion due to the existence of Jupiter would, at a distance of 4.3 light years, be quivalent to detecting a movement of 1/20th of an *inch* at a distance of 40 miles! If our imaginary Alpha Centaurian astronomers have telescopes similar to our own then they, like us, are incapable of detecting so small an element of deviation. To them our Sun would appear planetless despite its slow rate of rotation and they would remain in ignorance of the magnificent planetary family that orbits Sol. In like vein, slow rotating stars of the appropriate type and age which, to us, exhibit *no* proper motion deviation, could conceivably harbor solar systems as grand or grander than our own.

The only results available to date in respect to extra-solar planet detection are those due to astrometric measurements. The procedure is as follows. The star under examination is observed over a long period of time, its position being measured in relation to that of nearby "fixed" stars. Any irregularities in its proper motion are then analyzed to ascertain whether or not an unseen companion could be exerting a perturbing effect. This technique has now been in vogue for about four decades, but it must be admitted that many of the early results were of rather dubious value due to inaccuracies in the measurements. Indeed it must be admitted that even some of the more recent results are not above suspicion.

The stars that have been examined by this method can be conveniently grouped into two categories—stars for which the derived data may be spurious, and stars for which the evidence of the existence of unseen companions is good. Into the former category go Proxima Centauri, Luytens 726-8, Epsilon Eridani (oddly perhaps, since this star was one of the two selected for Project Ozma back in 1960), Kruger 60A, 70 Ophiuchi. The second, or favorable category, includes Barnard's Star (with which we will be dealing at length in the next chapter), Lalande 21185, and 61 Cygni, among others. In view of the inherent difficulties and complexities, however, it cannot be claimed that the results in respect to the second group can be endorsed categorically at the present time.

Apart from difficulties in astrometric observations, it must also be conceded that the accurate reduction of astrometric data brings problems. There are other problems too. Recently it was discovered that small astrometric changes relating to an M3 dwarf star, known as AC + 65° 6955, 27 light years distant, were due *not* to the perturbing effects of a planet or planets orbiting that star. Indeed there were no astrometric changes at all. The discrepancies were due to a fault in the objective lens of the twenty-four inch refractor telescope currently being used. Such are the hazards of astronomical research.

Astrometry, though of undoubted importance in the determination of parallel and of proper motion, is easily susceptible to errors and to misleading data reduction. Even when applied to the nearest stars, an element of doubt creeps in as to whether the observed perturbations in the motions of these stars can safely be attributed to the presence of unseen (planetary) companions.

It is becoming obvious therefore that another method of detecting extra-solar planets must be employed if definitive results are ever to be obtained. The use of a technique which could be so easily misleading is hardly acceptable, for on the results so very much depends. Is our solar system unique? Are there other solar systems? Are we alone in the universe, or do other cosmic civilizations abound? The questions are in so many respects fundamental. Stellar theory suggests that planets *are* likely around many stars and the results of astrometric observations would in many instances *appear* to bear this out. Nevertheless the situation is rendered tantalizing by the fact that the results of astrometric observations are not without dubiety.

It is now coming into the realms of possibility that extra-solar planets might be detected by photometric means using space-mounted telescopes of sufficiently large aperture. Two variations in the basic technique have so far been suggested. One is the construction in Earth orbit of a very high resolution optical interfermoter comprising an array of parabolic section mirrors with an optical resolution equivalent to that of 125 meters (approximately 500 inches). This method would probably permit *direct* observation of Jupiter-sized planets (or larger) of the very nearest stars, though it would be highly expensive to develop, construct and maintain. Alternatively a three-meter space telescope and an occulter positioned thousands of kilometers in front of the telescope might be employed. (For the record, an occulter is a device to *conceal* from view that which it is *not* desired to see.)

Recently a study along these lines was carried out by James Elliot of Cornell University (*Icarus 35,* p. 156) who proposes that a space-borne telescope, similar to the 2:4-meter instrument proposed by NASA for launching in the early 1980s, could use the *Moon* as an occulting edge in a search for planets in stellar systems up to about 33 light years distant—surely a novel role for our large natural satellite.

By precise adjustment of the exact size and shape of the space-vehicle's highly inclined Earth orbit, the orbital motion of the telescope could be matched with that of the Moon. This would then permit the dark lunar limb which is unilluminated by either Earth or Sun, to be aligned with great precision so as *just* to cover a star under examination, thus leaving planets a mere fraction of a second of arc to one side of the fully visible limb.

A "stationary" Moon as viewed from the telescope-bearing spacecraft could, it is reckoned, be maintained for approximately two hours though only some twenty minutes or so would be necessary to detect a Jupiter-type planet at a distance of 33 light years. The planet, it is calculated, would be some 300 million times fainter than its parent star, but the distant and airless lunar limb would provide a near perfect occulting edge. Consistent and repeated viewing of each prospective star over a period of several years would embrace all likely orbits for giant planets out to tens of astronomical units.

Use of the lunar limb in this manner would alleviate the highly critical alignment problems which would almost certainly be encountered were an artificial occulting edge to be used, since an artificial device would of necessity be so very much smaller and closer to the instrument-bearing space craft.

Another distinct advantage of a spacecraft in a high orbit would lie in its potential for recording slow lunar occulations of virtually any star in the sky with resolutions of at least one-thousandth of an arc second in the direction of motion. This would, it is believed, provide accurate structural details of many stellar discs, binary stars, quasars and pulsars. A one- to three-meter telescope, designed and launched specifically for such work, would probably prove one of the most productive and valuable scientific satellites ever placed in orbit. But whether such a satellite intended for purely scientific work would receive the necessary funding priority remains today especially questionable. To those politicians and others responsible for this aspect, we can only say that the scheme would lead to the acquisition of knowledge. Failure to carry it through means lack of knowledge—which is generally defined as ignorance.

During 1967 a novel idea was conceived by a number of American astronomers, physicists, and engineers to which was given the title of "Project Orion." This, it is felt, represents a cheaper method whereby extra-solar planets could be detected. It involves a new type of telescope capable of detecting planets orbiting other stars out to 32 light years from the Earth. This would cover the four hundred nearest stars and perhaps identify the most likely sites for terrestrial type planets and likely destinations for future interstellar probes.

The scientists responsible for Project Orion, and to whom due credit must be given, are Dr. David C. Clark (Project Director), NASA Ames Research Center; Dr. George Gatewood, Allegheny Observatory, University of Pittsburgh; Dr. Charles Kenknight, Optical Sciences Center, University of Arizona; Dr. Krzysztof Serkowske, Planetary Sciences Department, University of Arizona; Dr. Onestes Stavroudis, Optical Sciences Center, University of Arizona.

The telescope envisaged would embody what is known as an "imaging interferometer." If the Earth were devoid of atmosphere every star seen through a telescope would have the appearance of a small, bright steady disc (since we require an atmosphere if we are to live, it is clear we cannot have it both ways). Our planet has a very thick atmosphere. Indeed it is more than thick, it is decidedly turbulent. The direct consequence of this is that when we focus a small telescope or pair of binoculars on a star it oscillates rapidly (which brings us quickly back to the "twinkle, twinkle little star" nursery rhyme of our early childhood days!). The effect is very beautiful. Try a small telescope on Sirius or Rigel on a clear, dark and moonless frosty night. The effect may be very lovely but to astronomers it constitutes an unmitigated nuisance of the first order. Because of this oscillatory effect precise motion of a star relative to other stars is difficult to measure, the more so since oscillations of various stars due to the terrestrial atmosphere are independent of one another.

This situation can be improved upon if *two* telescopes are used in conjunction to form a *single* image of a star field. This, then is the principle on which the imaging interferometer is based. The two telescopes each have a single movable flat mirror and two stationary concave mirrors enclosed in a vacuum.

Received stellar light from a pair of linked telescopes does one or the other of two things. Either they add together *in phase* thus doubling the unwanted effect or they interfere and cancel each other out. The image of a star resulting from a pair of linked instruments is a round disc which is crossed by a number of bright and dark, rainbow colored parallel fringes. The central bright fringe is white and defines the *precise position* of the star (in a horizontal direction) and is thereby virtually unaffected by the terrestrial atmosphere.

This central white fringe is located at the precise point at which the lengths of *two* light paths from a *single* star through the *two* linked telescopes are equal to within one tenth of a light wave. This is extremely close, for the length of light waves from crest to crest is exceedingly short (we must remember the extremely high frequencies of light waves compared to most other forms of electromagnetic radiation, e.g., radio waves).

At this juncture we might care to consider a hypothetical example. Suppose another star, very similar in all respects to our own Sun, exists about ten light years from us, i.e., roughly 60 billion miles. Suppose also that among the family of planets orbiting this star there is a single, very massive planet akin in most respects to our giant Jupiter. As this planet moves steadily in its orbit around the star, the star develops a "wobble" around the center of mass of the planetary system. This wobble, however, is no tremendous deviation. It amounts only to one *billionth* of the distance from us. Clearly this is very slight, not at all obvious and seemingly impossible to detect.

Here then in stark, practical terms is a measure of the problem, but with the arrangement devised by Dr. Clark and his team the problem is greatly simplified. The separation between the *flat* mirrors of the two telescopes must be a *billion times* greater than the accuracy used to measure the position of the white center fringe (formed by combining the star light with the two telescopes). The position of this white fringe must be measured to within *a tenth of a wavelength.* Thus separation between the two instruments must be equal to 100 million waves—in practical terms a distance of sixty yards. By increasing the separation between the instruments the precision of the measurements may be augmented.

In point of fact *two* such interferometers are necessary. The first measures the *horizontal* component of separation between two stars when they are located, say, in the southeast. A few hours are then allowed to elapse until, due to Earth's rotation on its axis, these same stars lie in the southwest. The second interferometer is then used, which once again measures the *horizontal* component of separation between the two stars. However this second horizontal is in effect at *right angles* to the first, i.e., it is actually a *vertical* component with respect to the first.

We thus have a method for determining accurately the separation between two stars, and any very slight variations in the distance separating them is indicative of a slight regular deviation in the proper motion of the star suspected of having planets—a deviation which, for the reasons just stated, would not normally be discernible. This would point to the *possibility* of planetary presence.

Irrespective of all such schemes, however, there can never be much doubt that the positive existence of planets orbiting the nearer stars will always be difficult to determine from the surface of our planet. Only out in space, well removed from the deep and ofttimes turbulent atmosphere of Earth, do the best chances seem to exist, and even there the difficulties and complexities will be very real. In time we should be able to dispatch planet-seeking telescope-mounted probes to the very fringes of the Solar System. This should be advantageous, but even probes of this nature located some little way beyond the orbit of Pluto would still be exceedingly remote from the nearest star to the Sun, Alpha Centauri. A probe in the vicinity of Pluto lies over five and a half *light hours* from us (roughly 3,700 million miles). Alpha Centauri is 4.3 *light years* distant (roughly 25 million million miles). Thus any advantage gained is infinitesimally low. Probes going out well beyond the orbit of Pluto would assist, but the advantages gained would still be relatively slight. At times it almost seems as if the only way to be sure about our "cousin" planets is to go and take a look. But then of course, there are a few slight problems with that, too.

It has been suggested that in some future era it might be possible to achieve our aim by electronic means, e.g., sensing very high frequency radio or laser beams. This is clearly something well beyond the present state of our technology. Indeed application of such a technique at ranges of an interstellar order may never turn out to be practical. The time factor also enters the picture. A sensing beam looking for planets around Tau Ceti is going to take eleven years to get there and then having somehow impinged upon and revealed any planets, the beam, now a reflected one, is going to take another eleven years to get back to us. In the case of nearer stars the position is better, although it would still take about nine years to electronically locate planets about Alpha Centauri.

$$\textbf{6}$$

$\mathcal{W}\varepsilon\ \mathcal{A}$ʀε $\mathcal{N}$ot $\mathcal{A}$ᴌοnε

T HE TITLE OF this chapter does not imply, at least at this stage, that alien life forms and alien civilizations exist. What it does imply is that our *solar system* is not alone and is therefore not unique. In the previous chapter we dealt at some length with the detection of extra-solar planets and the complexities and difficulties inherent in such an undertaking. In the process we mentioned one or two distinct possibilities, e.g., 61 Cygni and 70 Ophiuchi, though it must be admitted that an element of uncertainty still hangs over these and certain others.

In this chapter it is fitting to dwell at some length with a stellar planetary system which, in the light of contemporary evidence, can be regarded as reasonably well confirmed. The star concerned is the well-known Barnard's Star which lies only six light years from our Sun and planets. By interstellar standards this is very close. The planetary retinue of this star can therefore probably be regarded as the nearest other solar system to our own. It must be emphasized, of course, that Barnard's Star is *not* the nearest star to our Sun. This distinction belongs to the Alpha Centauri/Proxima Centauri system some 4.3 light years remote. This is a triple star system (or perhaps more accurately a double star or binary system with a far-out third member). We will be dealing with multiple stellar systems in a later chapter since the implications for planetary orbits in such cases are involved and less promising.

It was in the early summer of 1963 that Dr. Peter Van de Kamp, Director of the Sproul Observatory in Pennsylvania, announced the discovery of a planetary companion to this small and insignificant red dwarf star which is presently the second nearest star to our own system. Since then, after further extensive observational work, the presence of *two* unseen planetary bodies was indicated.

This star was discovered in 1916 by the famous American astronomer E. E. Barnard. It is a single entity of magnitude + 13. It is therefore well beyond the limits of naked-eye observation. (The brighter the star the lower its magnitude. Very bright stars are said to be of first magnitude and on a clear, moonless night the unaided eye can discern stars down to sixth magnitude. A star of magnitude 13 would require at least an eight-inch telescope.) Spectroscopic examination shows it to be of spectral class M5 with a surface temperature of 2,900°K. It is, in fact, a small, red dwarf star, one of the most common types in the universe. Eddington later assessed its diameter as about 0.15 million miles and its mass as only 15 percent that of the Sun. It is not, therefore, a very impressive star.

At that time, apart from being recognized as the second closest star to the Sun, Barnard's Star was also recognized as possessing a further claim to distinction, for it possessed the largest proper motion of any known star (10.25 seconds of arc per year). All stars have a proper motion, i.e., they move relative to one another. The fact that from night to night, year to year, century to century and more, stars remain fixed, seemingly immobile in their respective constellations, is due merely to their immense distance from us. The northern circumpolar constellation of the Plough or Big Dipper is a classic example. Figure 3 shows us the constellation as it is today, as it appeared 200,000 years ago and as it will look to our distant descendants some 200,000 years hence. Even the eternal stars are not unchanging after all.

Barnard's Star is close to us and is at the same time a real hustler. Its 10.25 seconds of arc per year may not seem very much but, astronomically speaking, it is considerable. So, though at present the second closest star to us in the heavens, it will not

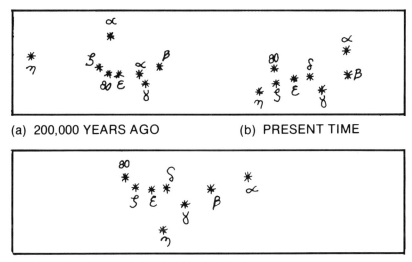

(a) 200,000 YEARS AGO

(b) PRESENT TIME

(c) 200,000 YEARS HENCE

Figure 3

retain this status for any great period of time—again, that is, by astronomical standards. Because of this rather unique feature Barnard's Star has earned itself the title of "Greyhound of the Skies." At the time of its discovery it seemed devoid of any other unusual or distinctive properties.

In April 1963 however, Van de Kamp intimated that he had detected a minute twenty-four year cyclic perturbation in the proper motion of the star. In other words the star was not cleaving a straight path through space. Superimposed on its normal straight line of passage was a regular fluctuation. In short, it was wobbling as it went along. As we explained in an earlier chapter, such movement can be attributed to the presence of an unseen, orbiting companion (or companions). From the degree of perturbation and the assumed mass of the star, Van de Kamp estimated the planetary mass as being 1.5 times that of our own planet Jupiter. This companion body is apparently moving in an elliptical orbit at a distance of approximately 400 million miles from the star. Since such a mass is almost certainly too small to initiate or sustain thermonuclear reactions it must be classed as a planet, albeit a fairly large one.

Earlier research which had indicated the presence of companions to the stars 61 Cygni 70 Ophiuchi, Lalande 21185 and Ross 614 indicated the masses of these to be about eight to ten times that of Jupiter. Consequently it might be more fitting to regard such bodies as "sub-stars" rather than as very large planets. In a "sub-star" certain nuclear transitions might be occurring and this would cause the body to glow. In fact, in the case of the star Ross 614, the companion *has* radiated sufficiently to show up as a faint photographic image in its predicted position. We can be quite certain that this is in no way due to reflected light from Ross 614.

Van de Kamp based his findings on an analysis of no less than 2413 photographic plates exposed on 619 nights from 1938 onward using a twenty-four inch refracting telescope. The resultant graph (Figure 4) illustrates rather well the cyclic nature of

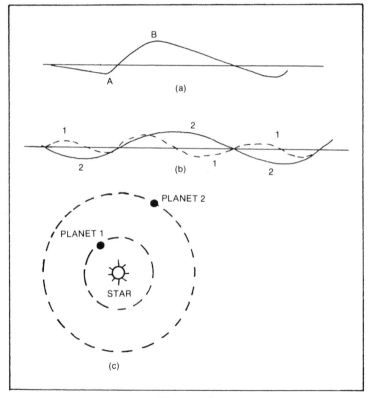

Figure 4

the perturbations. It should be pointed out of course, that the plotted values on this graph did not all lie neatly along the curve.

The latter represents the resultant or "best fit" curve from the experimentally plotted positions. The curve has, in fact, been "smoothed," a process that has to be done with the utmost care and discretion. It should also be borne in mind that such observations are by no means easy to make, and a margin of error is always possible. This can best be reduced, and the accuracy of the curve thereby enhanced, by the inclusion of further points on the graph, i.e., by further observations.

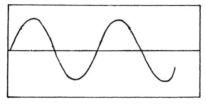

Figure 5

The curve is not what mathematicians would term as sinusoidal. An example of a sinusoidal curve is shown in Figure 5. This is regular and the amplitude or height of each peak is the same, as well as the period of fluctuation. The curve, as it stands in Figure 4 is said to possess a number of harmonics (one or probably two). These are really further curves superimposed on the original and have the overall effect of distorting the curve. This might be attributed either to the companion body moving in a highly eccentric, elliptical orbit *or* the presence of *more* than one companion. It is the effect of the first alternative which is really shown in Figure 6, that of the second in Figure 4.

In June 1969, Van de Kamp announced the result of further work. This confirmed the perturbations observed earlier, but also indicated a more elongated orbit and more definite signs of secondary perturbation. What all this adds up to is that the irregularities in the proper motion of Barnard's Star can be explained by the presence of *two* companion bodies, designated B1 and B2. Both of these are thought to orbit in roughly the same plane, the former having a mass 0.8 times that of Jupiter and describing an almost circular orbit of radius 258 million miles. The orbital period or "year" is put at twelve terrestrial years. B2 is reckoned to have a mass 1.1 times that of Jupiter and to describe a near

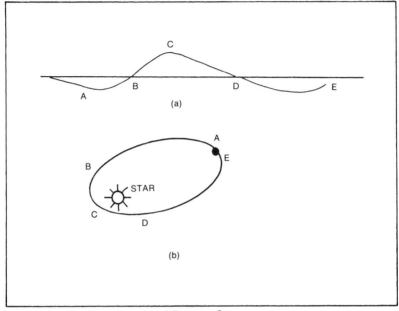

Figure 6

The curve at (a) can be explained by one companion moving in an eccentric elliptical orbit as at (b). The curve sections A, B, C, etc., correspond to companion positions A, B, C, etc.

circular orbit of about 400 million miles radius. In this case the orbital period is approximately equal to twenty-six of our years.

On the evidence it seems fairly reasonable to assume that B1 and B2 can be regarded as true planets. B1 would appear to be intermediate in mass between Jupiter and Saturn with B2 having a mass slightly greater than that of Jupiter. It is interesting at this point to compare the position with that of the Solar System. Are B1 and B2 in fact analogous to Saturn and Jupiter in our own system? Our system has also several minor members (e.g., Mercury, Venus, Earth, and Mars). Is this also true in the case of the Barnard Star system? It is, as yet, impossible to answer this question since the additional perturbation harmonics due to small low mass planets are impossible to detect. (Astronomers on a world of Alpha Centauri 4.3 light years distant might just be able to detect the presence of Jupiter and Saturn, perhaps even that of Uranus and of Neptune by this method but, unless they had been able to devise very sophisticated techniques, they would remain totally unaware of the presence of Mercury, Venus, Earth, and Mars.)

Accepted theory relating to the formation of the Solar System maintains that both composition and mass of the small and of the giant planets are due, not to chance, but to the materials remaining after the formation of the Sun. These materials subsequently condensed. First to do so were the metallic and rocky materials. Liquids such as water and ammonia condensed later. The former gave rise to the small solid planets such as Earth and Mars, the latter to the "gas giants" Jupiter, Saturn, Uranus, and Neptune. We may therefore justifiably speculate whether analogous processes in the environs of Barnard's Star produced not only its "gas giants" B1 and B2 but also several small "terrestrial" type planets more akin to Earth and Mars.

There is also a theory (due to Lyttleton) which claims that Earth, Moon, and Mars were formed more or less simultaneously. Should this be so it is conceivable that Barnard's Star *could* have three small inner planets followed by outer "gas-giants" B1 and B2. The fact remains however, that there is little likelihood of establishing the presence of such small planets at so great a distance.

It is equally feasible that the primordial nebula or dust-cloud responsible for the formation of Barnard's Star and its attendant solar system was of considerably smaller dimensions than the one which gave rise to our Sun and its system of planets. Thus there might have been insufficient material to create more than say half a dozen companions (i.e., planets) at most.

The possibility of a life-supporting planet orbiting Barnard's Star is so slight as to be virtually non-existent—at least, that is, in the sense of carbon-oxygen based life which, up until now, is the only one of which we have either knowledge or experience. From Figure 7 it is apparent that terrestrial-type planets (if any) lie much closer to their central star than do the inner planets of Earth. This is due to the fact that Barnard's Star is much smaller than our Sun. As a consequence the radiation is much less, only in fact, about 0.0001 that of the Sun. It is almost certain therefore that matter of a specific type will condense nearer to the star than

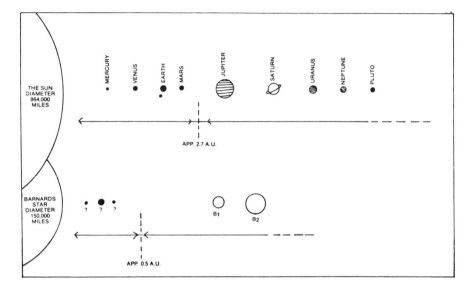

Figure 7

it would in the case of the Sun. This trend is substantiated by the fact that B1, in all probability a "gas-giant" like Jupiter or Saturn, lies at 2.8 A.U.s (approximately 260 million miles) from the star. This is much closer to the central luminary than is Jupiter in relation to the Sun.

Due to reduced radiation the concentric shell of space surrounding the star appropriate to the initiation, development and continued existence of life (the ecosphere) is very narrow. Outside this critical region temperatures are either too high or too low for biological processes to take place. The ecosphere with respect to our Sun is very much more extensive and corresponds approximately to the region between the orbits of Venus and of Mars. Since Venus has been found to be considerably hotter than we originally believed, our ecosphere should probably commence somewhere between the orbits of Venus and of Earth. The high temperature of Venus, however, is attributable in considerable measure to the retention of heat due to its carbon dioxide atmosphere (the famous "greenhouse effect").

So feeble is the radiation from Barnard's Star that the outer limit of its ecosphere (the coldest part) is less than the orbit of Mercury. Since on Mercury the withering torrent of heat, light and other radiations from the Sun is so great that molten pools of lead could collect in places on its surface (assuming lead exists on Mercury), we can discern the very fundamental difference between the various types of star and their respective biological potentials. For a planet orbiting Barnard's Star to receive radiation in intensity equal to that received on Earth from the Sun it would have to lie only about three million miles from the star. This is sufficiently close for the effect known as captured rotation to manifest itself, i.e., that planet's day and year would be of equal length; or, in more homely terms, were this to apply to Earth the day would start on January 1 and end on December 31. This, of course, is due to the planet turning only once on its axis during one revolution of its central star.

There is a further complication as far as life processes are concerned. Barnard's Star is a red dwarf. Such stars are prone to flaring, that is, there is at intervals a sudden increase in light and radiation intensity of several magnitudes over a very short period of time. This should not be confused with true stellar variability, a subject with which we shall be dealing in a later chapter. Our own Sun which is not, of course, a red dwarf but a class G yellow star on the main sequence, is also prone to this effect. When it occurs we say that a solar flare has taken place. Due to Earth's distance from the Sun (92 million miles on average) and its thick atmosphere solar flares are virtually harmless to us, though on the surface of Mercury (33 million miles from the Sun and with no atmosphere) they would be highly lethal.

Life, so far as we know, starts in oceans which in time become a sort of primordial organic "soup." Intense radiation from flaring would almost certainly destroy the cells of all initial organisms. The *nearest star* to the Solar System, Proxima Centauri, is also a red dwarf, so all that has been said in respect of the more distant Barnard's Star is true in this case also. Proxima Centauri, however, is merely the third and undistinguished remote member (if indeed it is a true member and not just a very near neighbor) of the binary (twin) Alpha Centauri system. The two components proper of the latter might prove more favorable to life. This aspect will be examined in greater depth when we come to deal with double and multiple star systems in a subsequent chapter.

In closing we can say with a reasonable degree of certainty that the second nearest star to the Sun has almost certainly a planetary system of sorts but the possibility of even lowly life there is virtually non-existent. The existence of another planetary system serves to indicate that the Sun is not unique in this respect. If more than one solar system exists, then it is reasonable to assume that the process leading to them could be at work in many parts of every galaxy. If we, or at least our descendents, ever have to flee the Solar System, there should exist havens for them.

(7)

Multiple Suns

THE STAR-POWDERED panoply of the heavens on a clear, moonless, winter night is without doubt one of the most magnificent spectacles that nature can provide. The realization that each
little point of light is a sun in its own right does much to com-
pound the sense of reverence and awe. But although naked-eye
observation certainly does not reveal the fact, a very high proportion of stars are actually double or multiple systems, i.e., systems
of two or more stars or biting common centers of gravity. In many
instances a small, relatively low power telescope or even a pair of
binoculars will reveal some of these stars for what they really are.
And very lovely the effect can be too, since in so many cases there
are distinct differences in the colors of the component stars, e.g.,
blue and white, red and yellow, green and orange, amethyst and
pink. Seen against the intense, dark background of the night sky
they appear in the field of a telescope or binoculars as bright
jewels exquisitely set in black velvet. Their sheer beauty is something of which no true lover and watcher of the night sky ever
tires. A very well-known example is the red super giant star
Antares (Alpha Scorpii). This is the principal star of the constellation Scorpio, the Scorpion which sprawls upwards from the
southern horizon during the long, warm semi-twilight nights of
mid-summer. The red Antares is found with optical aid to possess

a lovely green companion star. Indeed the two have been described as "celestial ruby and celestial emerald set in nocturnal velvet."

So far as the essential theme of this book is concerned the main interest regarding binary and multiple star systems is whether or not planets are likely to exist within them. It is this question to which we will now turn our attentions. It is a particularly interesting one.

In stellar systems of this nature two or more stars move around one another by virtue of mutual gravitational attraction. More correctly they both revolve around a common center of gravity just as our Earth-Moon system does, though in the latter case the center of gravity of the system lies *within* the Earth itself—an intriguing thought! Most common are the binary or double star systems. These particular systems fall into two general categories: very close binaries whose periods of revolution can be measured in days, and wide binaries the periods of which must be measured in years. It is uncertain at present whether the two types are distinct, the result of differing origins, or whether they merely represent subdivisions of a general population.

Multiple systems tend on the whole to be somewhat less common. In a certain number of instances the term "multiple binary" could be regarded as the more accurate despite the apparent contradiction in terms. It is, for example, possible in a triple system for two of the components to lie relatively close to one another with the third member very much further out. Probably the best known example of this is provided by our nearest stellar neighbor, Alpha Centauri. The two principal and relatively close components in this instance are Alpha Centauri A and Alpha Centauri B. One is a type G star closely akin to the Sun whereas the other is of type K and somewhat cooler. The two lie 24 astronomical units apart (over 2,000 million miles) and have a revolution period of 80 years. The third member of the system is a "cool" class M red dwarf star which apparently orbits the other two at a distance of 1/10 of a light year. It is presently closer to us than its two brighter companions and has for this reason been designated as Proxima Centauri. Doubts are sometimes expressed as to whether Proxima Centauri is really a bona fide member of the Alpha Centauri system on account of the distance separating it from the other two. At the present time the consensus would seem to be that it is.

A quadruple multiple system can come even closer to being a "multiple binary" if the four stars comprising it are grouped in two sets of *widely separated pairs* (Figure 8a). An alternative arrangement could be a "close" binary pair, a moderately distant third component and a very distant fourth component (Figure 8b). In the case of systems having more than four stellar components the number of possible combinations merely increases. The same principles are retained. The idea of a star system having more than four components may seem somewhat bizarre. Nevertheless such systems *are* known to exist and indeed one can be seen very easily during any clear winter night in the northern hemisphere (though, of course, only as a single star to the naked eye!). This is the bright star Castor in the constellation of Gemini, the Twins, a short distance to the east of the red star Betelgeuse in Orion. Castor is in fact a magnificent system of no fewer than six separate, but physically related stars.

It is fair to state that the majority of stars represent binary or multiple systems. Indeed it would appear that stars rarely come into being as single entities and in this respect our single Sun, though certainly not unique, is undoubtedly one of a minority group. Of course it has been suggested that this was not always so, that once our Sun had a binary twin which was overtaken by some sort of cosmic disaster. There is however no evidence for such a belief, which can only be regarded as the sheerest speculation.

The formation of stars from primordial clouds of hydrogen and dust in galaxies seems to lead almost as a matter of course to the formation of multiple or binary star systems. In the case of our Sun and other solitary stars it is thought that a somewhat analogous process takes place, except that in such instances what is spawned is not a number of stars but a single star and a number of *planets* and lesser bodies. The question then arises as to whether planets form coincidentally with multiple stars also. This is clearly highly pertinent to our theme. Present thinking is that multiple and binary systems *could* have planets.

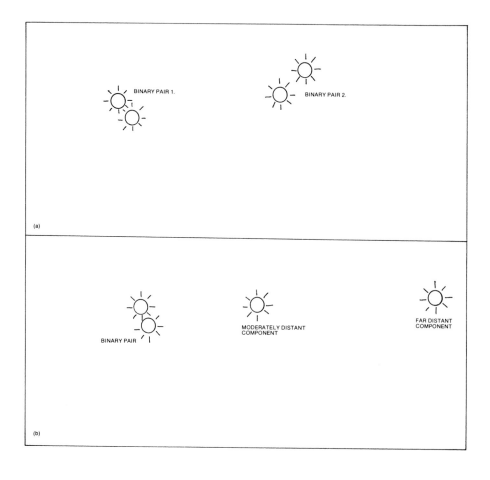

Figure 8

Assuming then that this is so, the matter of planetary orbits in such systems becomes of considerable interest and importance. Of even greater consequence would be the *stability* of such orbits. Would these become increasingly eccentric with time?

Planets of binary or multiple star systems could have complex orbits, depending on the relative masses and separation of the components. In some of the possible orbits the distances separating planets from their respective suns could vary greatly, leading to very extreme climatic conditions. Protracted periods of cold, with temperatures of an order low enough to solidify water, other liquids and many gases would certainly preclude the existence of active forms of life, though low order organisms *might* just survive. Biopoesis and evolution might proceed during periods of sufficient warmth, with organisms simply remaining dormant during the cold seasons. Conversely the closest approach of a planet to its sun might raise temperatures to such an extent that complex carbon compounds would break down. In these circumstances life could never emerge.

Let us however, look a little more closely at this particular aspect. In the first instance we will consider a planet in a simple binary star system (Figures 9, 10). The celestial mechanics in the case of a planet constitute the classic case of a three body problem. The two body problem, i.e., a single planet orbiting a single star, is easy to solve and merely invokes the famous elliptic law formulated by Johann Kepler. Unfortunately the situation in respect to a *three* body problem is considerably more complex and, in fact, there is no simple solution to it. In the past it was essential to make certain simplifying assumptions and hope for the best. This was unsatisfactory in diverse ways and hardly appealed to astronomers. In the prevailing circumstances of the period there was little that could be done about it. The advent of the computer, however, rendered possible an entirely new approach to the problem. This technique was based on the use of a number of numerical experiments and by this means it became feasible to follow the motion of a particular system, determining whether or not that system could be stable. If this is done in a sufficiently large number of cases, varying the parameters for each, it becomes

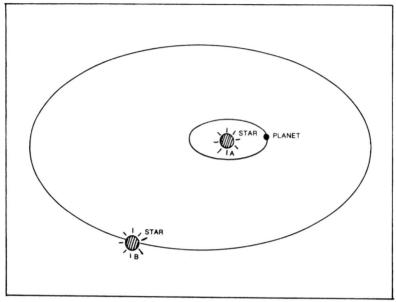

Figure 9

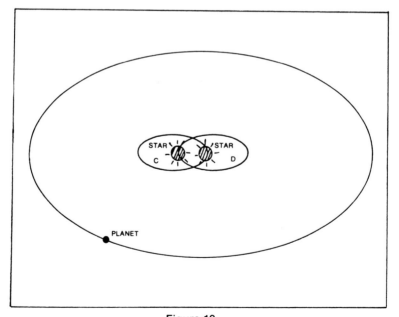

Figure 10

possible to establish the essential factors for stability and the limits of these factors.

The results of such computerized experiments have produced results fairly well in accord with those anticipated. Stable planetary orbits fall into two classes. The first of these (Figure 9) is an example of a planet orbiting one of the stars, the companion star in turn orbiting the former *and its planet.* Figure 10 on the other hand portrays a rather different set of circumstances, for here the planet is distant from both stars which are, of course, revolving around their common center of gravity. In this instance the planetary orbit is circular whereas the stellar orbits are elliptical.

How then does all this affect planetary stability? In the first instance (Figure 9) the planetary orbit will remain stable so long as the outer, more remote star does not approach within 3.5 times the distance of the planet from star A. In the second case (Figure 10) the orbit of the planet will remain stable so long as the planet remains more than 3.5 times the mean separation of the binary pair C. D. It will be seen therefore that regions in which stability is assured are fairly extensive. In other words, binary star systems *could* have planetary families though the conditions, notably with regard to orbits, are slightly more restrictive.

It would probably be of interest to consider for a few moments what the position would be here in the Solar System were our own familiar Sun one of a binary pair. Of course to do this we have to change the Solar System about quite a bit, the major piece of cosmic surgery being the replacement of the planet Jupiter by Sol's *hypothetical* companion star. This star, we will assume, has virtually the same mass as that of the Sun. Immediately we have a very different Solar System. What happens to the planets must be the inevitable question. So far as the inner planets are concerned, Mars, at once, has a very rought time. Its motion is immediately and seriously perturbed and this causes it to wander erratically from near Earth's orbit to a point out well beyond the asteroid belt. Indeed it is possible that it might in the end escape from the Solar System altogether thereupon to wander aimlessly and eternally in the black unending night of interstellar space. It is however, believed that in the same circumstances Earth's orbit would remain stable by virtue of the fact that Jupiter's position in the Solar System (and by definition that of Sol's companion star) is 5.2 astronomical units (over 473 million miles) from the Sun. But though remaining stable, there might be a slight variation in Earth's orbit. It is exceedingly unlikely however that this would cause any marked physical manifestations.

If now we try the same "experiment" in the case of the second possibility the position becomes rather different. In this instance we are "replacing" our Sun by a *pair* of suns 0.2 astronomical units (about 18.5 million miles) apart. These have a revolution period of 33 days. In these circumstances the planet Mercury is immediately expelled from the Solar System to follow the example of Mars in the preceding case. The orbit of Venus would, it is believed, remain roughly similar to what it is at present, though it might show an enhanced eccentricity. Once again Earth's orbit could be expected to remain essentially stable, its distance from the twin suns being five times the mean binary separation, i.e., one astronomical unit or 93 million miles. Thus, in binary star systems of the two kinds we have been considering, a planet like Earth, located in a like orbit, would probably be just as tolerable to life forms on its surface, so long, of course, as the stars concerned were of solar or very similar type.

Let us now consider the case of the nearest star to the Sun, Alpha Centauri. In chapter 6 we dwelt at some length on the planetary system of Barnard's Star, one which we now have really good reason to believe does exist. As yet we have no such evidence for the existence of a planetary system in respect to Alpha Centauri and in this instance we are merely speculating on *possible* conditions should a planetary retinue really exist there. One of the two principal stars in the Alpha Centauri system is of type G, i.e., it is very similar to our Sun in most respects though it may be just a little older. Its companion is of type K, slightly cooler than its solar-type "twin," and therefore of course slightly cooler than our Sun also. The two stars are 24 astronomical units apart (about 2180 million miles), the revolution period around their common center of gravity being 80 years. Thus a planet one astronomical unit (93 million miles) from the G type (solar) star would be similarly placed to our Earth which also lies at this distance from a G type star.

But what, the reader may well ask, is going to be the effect of the K type star on that planet? The answer, it would appear, is very little indeed, since the planet in question lies well outside the distance limits for perturbation of orbit given earlier, i.e., more than 3.5 times. At a distance of 23 astronomical units from the K type star this condition is fulfilled many times over. Indeed there is no reason why *each* of the two stars should not have planets in orbit around them. Since one star, the G type component, is very closely akin to our Sun and the other, the K type, similart though cooler, there could be at least one planet in the habitable zone of each star capable of sustaining life. That orbiting the G type star would require to be approximately one astronomical unit from it, that orbiting the slightly cooler K type star about 0.6 of an astronomical unit. Thus, though the planetary system of Barnard's Star (six light years distant) is the nearest reasonably proven one to our own, it just might not, after all, be the closest. Our nearest stellar neighbor, Alpha Centauri, *could* also provide our nearest family of "cousin" worlds. As we have seen, Barnard's Star being a cool class M type dwarf could hardly provide the appropriate locale and environment for biological initiation and development. No such inhibitions apply so far as the Alpha Centauri system is concerned. We must again stress however, that, to date there exists *none* of the indirect sort of evidence for the existence of planetary dimension bodies there which we discussed earlier. Nevertheless we must always remember that twin (or multiple) star systems by their very nature are bound to possess an inherent "wobble" or deviation from their expected straight line trajectories. A further deviation due to planets, in effect a very minor secondary deviation superimposed on the greater, would be very hard to detect. Improved, much more sophisticated techniques may yet tell us of the planets around the twin suns of Alpha Centauri. In time we will probably have the answer.

The next closest multiple star system to us after Alpha Centauri is one designated by astronomers as L726-8. This is 8.6 light years distant. At present it is very difficult to say much about this particular system in respect to planets except that all biological connotations can be quite safely ignored. For life to have evolved in the environs of this particular star any planetary orbits would need to lie so close to the stars as to have highly unstable orbits—

and these we can safely assume would hardly be conducive to life. Even our science fiction friends would not be tempted to use this star and its environs as a locale for a space odyssey. One of the components is also a flare star (as is Barnard's Star). This too seriously downgrades life potential. So far as we know at present there is no indication of planets, but this does not of course rule out the possibility entirely.

Lying only 0.1 light years further out, at 8.7 light years from the Sun, is probably, indeed almost certainly, the most beautiful star in the entire firmament as viewed from Earth. So far as we are concerned it is also the brightest. Its name is Sirius, the legendary "Dog Star." There must be few watchers of the night sky who have not admired, during the frosty nights of midwinter in the northern hemisphere, this brilliant bluish-white star as it flashes, silent and resplendent, just above the southern horizon only a short distance to the east of the magnificent constellation of Orion. Though the color of Sirius is indeed bluish-white, its position low in the south means that its rays must penetrate more of the Earth's turbulent atmosphere than those of stars nearer to the zenith. Thus, and especially on frosty nights, Sirius, due to enhanced refractive effects, is capable of showing brief flashes of most of the colors in the spectrum. This adds very greatly to the sheer loveliness of the star and to the general magnificence of the night sky in winter.

Sirius is also a member of a binary star system, though the truth of this was not appreciated until as late as the year 1862. The story of the discovery of the companion of Sirius is well worth recounting. It begins some eighteen years earlier in 1844 when the celebrated German astronomer Friedrich Bessel discovered that Sirius (as well as Procyon, the little Dog Star in Canis Minor) did *not* show a normal proper motion. In other words the "wobble," which we have already described in relation to large planetary presence, was again manifest but in a much more positive and decided way (as could confidently be expected when a star rather than merely a large planet is responsible for the fluctuations). At this point Bessel had no choice but to leave the matter since no telescope then in existence seemed capable of revealing the presence of the perturbing body.

It so happened however, (and, as it turned out, most fortuitously) that in 1862 an American telescope maker, one Alvan Clark, already highly renowned for the excellent refractor telescopes he produced, was testing a new eighteen-inch lens for the Dearborn Observatory. Quite by chance he was using Sirius as the object of his tests, probably on account of its brightness. At that point Clark was endeavoring to ascertain how soon the light emanating from Sirius became perceptible before the star itself came into view from behind the corner of a building. To his intense surprise he noticed the appearance of a very faint companion to Sirius fully three seconds prior to the "arrival" of Sirius itself. This represented a really most remarkable achievement since Sirius is some *ten thousand* times brighter than its diminutive companion. As an immediate consequence Sirius proper was thereafter designated Sirius A and its minute companion star as Sirius B. Facetiously they have since been dubbed "Dog Star and Pup"!

Sirius, the Dog Star, is a very bright type A main sequence star. The "Pup," on the other hand, is simply a degenerate white dwarf star, the sad relic of a once great and mighty sun. The pair represent rather well both stellar youth and stellar age. The two are separated by a distance of 20 astronomical units (1860 million miles), and the period of revolution around their common center of gravity is fifty years. So far as life possibilities are concerned the habitable zone around Sirius lies about 4.5 astronomical units (over 400 million miles) from the primary.

The possibility of planets swinging in orbit around either Sirius A or Sirius B is, however, very doubtful indeed. Sirius A (i.e., Sirius proper) is a young, exceedingly hot star showing a distinctly high rate of rotation. As we have already seen, this is probably indicative of a young star that has not, so far, spawned any planets. The case of Sirius B (the "Pup") is that of a star which has left the main sequence aeons ago, passed through the catastrophic red giant stage during which it must have swelled out and consumed any planets that had the misfortune to orbit it. This stage over, the star then condensed into a white dwarf. Only far out planets would have escaped, and they would very likely have been well scorched.

At a distance of 11.2 light years we come next to the binary system known as 61 Cygni in the constellation of Cygnus the

Swan (on account of its shape, known sometimes as the Northern Cross). This was one of the first stars to be suspected of having planetary companions. Recent detailed studies by certain Russian astronomers suggests that it is accompanied by three Jupiter-type planets. 61 Cygni consists of two stars of 0.6 solar masses revolving around one another with a period of approximately 720 years. Magnitudes of the two are 5.6 and 6.3 respectively. Mean separation between them is approximately 80 astronomical units (over 7,280 million miles). Thus any planet orbiting in a habitable zone around either star would certainly be stable in that orbit. At present this binary is rushing past the Solar System (in the astronomical sense) and across our sky, at 5.2 arc seconds per year. Both components are K class dwarf stars and though therefore somewhat cooler than our Sun they should still be able to provide the energy to initiate and nurture forms of life. Such planets would need to lie within approximately 0.8 of an astronomical unit (about 75 million miles) of one or other of the stars. Not that we could reasonably anticipate life on Jupiter-style planets, but if such large ones do exist then so also may smaller, much more appropriate ones, just as in the Solar System.

A. N. Deutsch and O. N. Orlova of Pulkovo Observatory in a recent issue of *Astronomichesky Zhurnal* report their findings of the star's motion using 418 photographic plates from Pulkovo and 966 from the United States Sproul Observatory. These together embrace the motions of 61 Cygni over the last hundred years. Following careful removal of effects attributable to the 720-year orbital motion of the binary, they claim the existence of significant perturbations in the movement of the star which they interpret as being due to the influence of several massive dark companions (i.e., large planets).

The conclusion reached by Deutsch and Orlova is that the double star system of 61 Cygni is accompanied by *three* giant planets having masses of seven, six, and eleven times that of our own Jupiter with orbital periods of some six, seven, and twelve years respectively. There exists some evidence that the seven-year period planet is bound to one star in the double system and the other two to the second. (Figure 11).

If these facts can be substantiated they will almost certainly lead to a much more intense interest in extra-solar planet hunt-

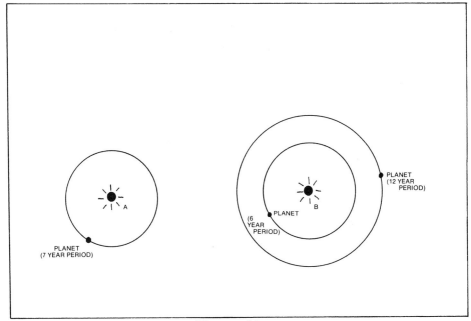

Figure 11

ing. This will be no bad thing. We can reflect on the possible results that might very profitably accrue if other near-by stars come under similar investigation during the next few years, and from the minute deflections in their motions that could be revealed.

Proceeding yet further out and away from the Sun we come to the bright winter star Procyon at a distance from us of 11.4 light years. Procyon—sometimes referred to as the "Little Dog Star" because it is the principal star in the constellation of Canis Minor, the Little Dog—was the other star which Bessel observed in 1844 to be deviating from its straight, proper motion path. As in the case of Sirius the inference of this was the existence of a companion star. We now know that the system resembles Sirius in another respect, comprising as it does a bright primary component and a dense white dwarf. Planets could be possible here, though life potential is regarded as very low.

Between 11.5 and 15.5 light years from the Sun there are six red dwarf binary systems, all of which could possess planets. All these stars are comparatively dim and cool and it would be hardly surprising were life in any guise or form to have gained a foothold around any of them. At a distance of 15.6 light years there are two *triple* systems, both in the constellation Eridanus. In each case one of the components is a white dwarf. So far it is impossible to say whether the existence of planets could be postulated here.

At 16.7 light years from Sun and Solar System we come to a most interesting binary system, that of 70 Ophiuchi. It was this star, along with 61 Cygni (mentioned earlier), which was first suspected of possessing a planetary companion or companions. Many works of that period tend to imply that the existence of a planetary family to 70 Ophiuchi is a well-proven fact. This is perhaps stating the position too strongly and too optimistically. The system comprises two K type stars, both therefore slightly cooler than our own Sun. The two component stars are 23 astronomical units apart (i.e., over 2,000 million miles) and have a revolution period around their common center of gravity of 88 years. The habitable zones are thus well within the stable orbit zones in the case of both stellar components. It must be emphasized however that the planetary body indicated this far would seem to be a very large one. Whether or not other smaller planets also swing around either or both stars remains an open question. Indeed it has been suggested that the true path deviation of the system is due to the interacting forces of the two stars as they swing about their common center of gravity and *not* to the presence of a large planet.

The star Eta Cassiopeia is currently regarded as a very strong possibility in the planetary stakes. Cassiopeia is, of course, the well-known north circumpolar constellation visible throughout the northern hemisphere all the year round. The star is 19.2 light years distant. One of the components is a GO main sequence star which means that it is virtually a replica of our Sun in respect to temperature, mass, and luminosity. The other component is a relatively dim class M red dwarf. The two orbit one another in a period of approximately 480 years, their mean separation being 70 astronomical units. The orbits are, however, eccentric. Nevertheless minimum distance between the two components is put at 35 astronomical units, which is roughly the distance between the Sun and its outermost planet, Pluto. The GO component is ideally suited in all respects to have a planetary retinue. Indeed a planet of similar constitution and size to that of the Earth at roughly 1 astronomical unit from the star would enjoy conditions comparable to those of our own world, especially if speed of rotation and axial tilt were close to 24 hours and 23½° respectively.

This would appear to represent a fairly promising region in any serious search for extra-terrestrial life forms. It could be asked of course, what the effect of the M red dwarf component might be. In fact, from such a planet this star would merely be an orange-red point of light. It would certainly be a very *brilliant* point of light and indeed rather a lovely object in the skies of such a planet. The magnitude, it is reckoned, would probably be of the order of –15 on average. We can profitably compare this with the light received here on Earth at night from a full Moon, which has a magnitude of –12.5. The red star in that planet's skies would therefore be around two and a half magnitudes *brighter* than our full Moon. This supra-brilliant stellar object would be almost stationary in the planet's sky, especially when it lay furthest away. During the night it would cast a sort of roseate twilight, and during that period of the year when it shone from the daytime sky it would be easily visible, though perhaps not particularly conspicuous.

It is an established fact that, as we proceed outward beyond 20 light years, the frequency of binary systems having a sufficient degree of separation to permit the existence of stable planetary orbits within ecospheric zones increases quite dramatically. In the fairly recent past it was widely accepted that multiple star systems were no place to anticipate the existence of extra-terrestrial life. Much was made of hypothetical planets that were alternately baked and frozen by reason of the stellar components engaging in a sort of cosmic tug-of-war with them. There even existed the rather incredible idea (it still rears its head from time to time) that it might be possible for a planet to pursue a *figure eight orbit,* swinging first around one planet then doing the same around the next, crossing its own orbit in the process (Figure 12). Were intelligent beings to inhabit such a planet they would find their day, their seasons, their year, indeed their whole lives to be most amazingly complicated. Figure eight orbits are best left to the realms of science fiction.

The position today is, that if planets do exist in binary systems, there is no valid reason why in many instances the orbits of some

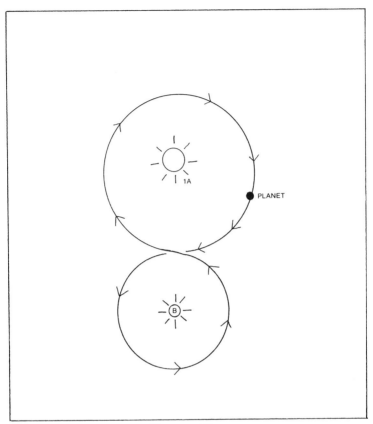

Figure 12

of these planets should not be perfectly stable. The orbits of others could possess slight to moderate degrees of eccentricity. Others again could have orbits eccentric in the extreme. At the same time there would appear to be a distinct lack of unanimity as to whether planets can exist in the environs of *multiple* stars. One body of opinion tends to regard their formation as highly unlikely in the circumstances, while another takes the diametrically opposed view. We mentioned earlier the belief that stars tend to form as physically connected adjacent bodies, i.e., binary or multiple systems, whereas single bodies such as our Sun have to content themselves with a mere planetary retinue by way of cosmic companions.

It seems rather unlikely however, that we can group these two alternatives in totally water-tight compartments. Stars, we know, form as a direct consequence of gravitationally condensing clouds of primordial hydrogen. If the cloud is sufficiently large then apparently condensation can occur at several centers within the cloud. The result is a multiple star system. Should the original hydrogen cloud be of smaller dimension, condensation takes place at one center only and a single star results. Any remaining material, for the most part, undergoes accretion to protoplanets and thence to planets. If this concept is valid there seems no reason why such accretion of residual material should not occur in the case of multiple star formations unless—and this seems rather improbable—*all* residual material goes into the stars themselves. If we extrapolate this hydrogen cloud—star formation the other way again, we might even account for the existence of open star clusters, e.g., the Pleiades. Here perhaps an extremely large gas cloud gave birth to upward of thirty stars. It may be of course, that this theory of formation is too neat and too facile. Perhaps the reason for just one or more stars forming is much more complex. What parameters determine the condensation start-point in the cloud? Could embryo condensing stars split up like amoeba in the process? Many questions are still unanswered.

Despite these doubts it is possible today for high speed electronic computers to model the collapse of an interstellar cloud of hydrogen gas under the influence of its own gravity. Whatever stages are passed through, it is now generally accepted that stars *do* form by virtue of such a process. Although these computer experiments can hardly at this stage be considered definitive or final, they do strongly suggest that when a star is born it is likely to possess a circumstellar cloud, halo, or nebula. Conditions within this cloud are probably highly conducive to the formation of planetary bodies. A single forming star would have this cloud and eventually its family of planets. Similarly two, three, or four stars (or more), forming at various points in one primordial gas cloud would each have a circumstellar cloud from which planets would eventually form. So long as the individually forming stars were reasonably spaced there would seem to be little likelihood of these circumstellar clouds interfering with one another.

The belief that newly formed stars possess a circumstellar disc of gas and dust has been substantiated recently. A faint star in the constellation Cygnus, the Swan, known as MWC 349, has apparently reached the stage where it is rapidly evolving an enveloping disc of gas and dust. This object might even be construed as the early genesis of a future planetary system. The central star contains the rough equivalent of thirty solar masses in a globe having a diameter about ten times that of the Sun. Spectral evidence suggests that the star is centered in a vast disc of dust and gas. The central portion of this disc is luminous out to twenty times the diameter of the star and apparently emits ten times as much light as the star itself. The total light output of the system is also reported to be declining at about 1 percent per month as nebular material spirals outward. Admittedly this star is a *single* entity and not a binary or multiple system. Nevertheless the evidence would at least seem to bear out the possibility of planet formation occurring in this way, a mechanism which seems as likely to be valid in the case of a binary or multiple system as in that of a single entity.

To sum up, planetary systems *can* exist. We ought to know since *we* exist in one. There seems no reason why others should not exist around other single stars. With binary and multiple star systems, conditions could be and probably are different. But as long as the stellar components constituting these systems are sufficiently far apart, planets should be able to orbit these components suffering little or no gravitational perturbing effects due to the physically-related but sufficiently remote sister sun(s). If the stellar components are close or very close then the position must be radically changed and such planets will of necessity pursue very peculiar and highly eccentric orbits. It could even be, under certain circumstances, that the gravitational tug-of-war exerted on such planets by two, three, four, or more suns could in time disrupt or break them up. Finally we must also consider the remote possibility that the existence of two or more gravitationally linked suns might, to some degree, inhibit the accretion or other process which leads to the formation of planets within such systems.

Before we leave the subject of binary/multiple stars and their possible solar systems it might be of interest to record an interesting and relevant suggestion made by E. R. Harrison in 1978. This speculates on the possibility that our Sun might *not* after all be a single entity! If such a theory has any validity it would mean that a binary system has definitely a planetary retinue, i.e., our own!

Harrison's premise is based on the fact that some pulsar observations could best be explained by the effect on the Sun's motion through space of a nearby companion, an otherwise undetected dwarf star, neutron star or black hole. A conventional star, a low mass white or black dwarf sufficiently close to produce the necessary orbital variations would be so bright in the infra-red portion of the spectrum that it would almost certainly have been detected by now. The options remaining open are therefore for a more exotic cosmic body such as a neutron star or a black hole. This possibility has been investigated by Serge Pineault at the University of British Columbia in Vancouver, Canada.

Obviously there are a number of problems inherent in such a radical idea. The most obvious is that if the Sun is really a member of a very loosely bound pair, then the supernova explosion by which the neutron star or black hole came into being should surely have disrupted the pair. In his investigations Pineault has confined his theoretical studies to models of a temporary binary system in which the Sun is briefly joined (by stellar standards) to a *wandering* neutron star or black hole brought about by a chance encounter.

Pineault claims that fourteen celestial X-ray sources are known which happen to lie more or less in the direction of the hypothetical companion (neutron stars and black holes are regarded as prolific sources of X-rays). Six of these lie close to the direction of the galactic center, but the eight others could have a better pedigree so far as being the Sun's companion is concerned.

For any of these sources to be a genuine companion to the Sun the object concerned would need to have a low mass (equal to that of the Sun or less). Were this not so, then it would be gathering matter to itself to such an extent that it would shine brightly (i.e., in the X-ray sense) at its presumed close distance. This distance is estimated at 800 astronomical units (about 75,000 million miles). These may seem very remote but it is still close compared to the nearest star to the Sun (Alpha Centauri), at a distance of 25 million, million miles (4.3 light years). Pineault states that the possibility of this body being a low mass neutron star is more valid than electing for one of mass greater than our own Sun, since this would probably collapse into a black hole.

118/WHERE WILL WE GO WHEN THE SUN DIES?

It is difficult to give very much credence to this idea, though undoubtedly it is highly interesting and extremely novel. A quote from *New Scientist* (issue of 26 October 1978) is worth giving here: "One thing is sure—if any of the eight candidates really is a companion to the Sun, its proper motion will make it stand out soon as it shifts position relative to the background stars. And photographic surveys of the error boxes of these eight sources might well be worth their weight in Nobel Prizes."

The collapse of a gas cloud to form star(s) and planets might, it has been suggested recently, be due to "triggering" by a nearby supernova. New isotope studies of the material in the Allende meteorite are believed to hint that this could be so. The concept of a supernova trigger for the Solar System is a most intriguing one—perhaps revolutionary would be a more appropriate adjective. The technicalities involved are, on the whole, too complex for a book of this nature. Investigation and research is still proceeding. These, to quote a common expression, are early days yet. Should the concept ever be shown to have validity it would radically change many of our ideas. Supernovae (as compared to ordinary novae) are extremely rare. If, therefore, stars and planetary systems owe their existence to them, then there should be few stars and planetary systems. We cannot at present be too categorical about other planetary systems, but clearly there are many stars. Indeed it has been said that there are more stars in all the galaxies of the universe than there are grains of sand on all the beaches and shores of Earth. This would mean rather a lot of supernovae.

If, then, for some reason in the remote future our Sun should render the Solar System uninhabitable, our distant descendents could perhaps find safety and security on a world or worlds owing allegiance to more than one sun. We might imagine a day in the future history of our kind when more than one sun rises in the morning sky, where fleeting double or triple shadows are normal, where sunlight could be yellow—and orange—and red!

Uncertain Suns

IN THIS CHAPTER our brief will be to consider the existence of planets orbiting stars of an unusual nature. The vest majority of stars, including our own Sun, are fairly stable bodies in respect of their output of light, heat, and other forms of radiation. So far as planets containing any forms of life are concerned this can only be regarded as an unmitigated and undisguised blessing. By and large the only extremes to which our Sun subjects us are the result of terrestrial geography. If we go to the tropics we find the Sun's heat very strong, though with reasonable safeguards and common sense, not unendurable. If, on the other hand, we decide for a sojourn in the Arctic or Antarctic we find the Sun's heat is at best minimal. These extremes are of our own making or at least of our own choice. We cannot hold the star we call the Sun responsible for them. Neither can we blame the Sun for the fact that our planet's axis is tilted at 23½° to the normal, thus giving us the temperature variations due to the seasons.

There are, however, a considerable number of stars in our galaxy that are not nearly so quiescent and well-behaved. Their output of heat, light, and deadly radiation can and does fluctuate very wildly indeed. These fluctuations may be regular over periods ranging from a few hours to several months, or they may be totally irregular and to all intents and purposes quite unpredictable. Such stars are known as variable stars.

Before dwelling on the planetary aspects, it might be advisable to take a closer look at these peculiar stars themselves. Although they are grouped under the convenient generic title of variable stars it is a recognized fact that such stars come in many guises. Once we have considered a few of these we will be better able to understand the implications for planets which have the misfortune to orbit any of them.

When we gaze up into the heavens on clear and moonless nights it might well appear that all the stars we see there maintain the same brightness from night to night, week to week, month to month, and indeed forever. For almost 2,000 years since the times Ptolemy and Hipparchus compiled their star catalogues this belief went virtually unchallenged. During the latter half of the sixteenth century, however, it was finally and irrevocably exploded with the realization that many stars vary greatly in their degree of brilliancy. The invention of the telescope and of the spectroscope, plus adoption of many ingenious observational techniques, have since enabled astronomers and astrophysicists to gain much more insight into the behavior of such stars.

Historically the story of variable stars is interesting. During the month of October 1572, there suddenly appeared in the constellation of Cassiopeia, a spectacular "new" star. This unexpected beacon in the celestial firmament was studied assiduously by the celebrated Danish astronomer, Tycho Brahe. At its brightest this remarkable and striking object was quite clearly visible in broad daylight. Within a few weeks, however, it had begun to fade and by the time its first anniversary had come around it had fallen to sixth magnitude and was thus only just visible to the naked eye on a clear and moonless night.

Twenty four years later, in 1596, Fabricius made the exciting discovery that a star in the constellation of Cetus, the Whale, showed distinct variability in its light output. Another astronomer of the period, Holwarda by name, observed the star assiduously over several years and was able by 1683 to state with some conviction that the star varied in brightness over a period of approximately 330 days. The star was named Mira (the Miraculous), a title it still retains. Later it was recognized that a high proportion of variable stars display light fluctuations of a similar character, i.e., a wide variation in brightness from maximum to minimum and periods from maximum to minimum in excess of ninety days. As a consequence such stars became known as long period variables.

In the early days of variable star searching, astronomers were greatly handicapped by the lack of suitable instruments. As a consequence the number of variable stars discovered remained rather low. About the middle of the nineteenth century however, Argelander began the first systematic search for these objects. This search was enthusiastically and ably continued by several other astronomers, amateur as well as professional. The technique was improved toward the close of the nineteenth century by the application of photography. Hitherto the standard procedure had been to compare star fields with charts prepared by other astronomers several years before. Stars not appearing in those earlier charts were then regarded as potential variables and observed closely for any change in brightness. With the advent of photography two plates covering the same portion of the sky could be examined under a blink microscope. In this way any stars the brightness of which had altered during the intervening period could be recognized fairly easily.

This technique has continued to the present day, and has by now reached such a state of sophistication that nearly 30,000 variable stars are known. This is a by no means inconsiderable number and more are continually being discovered.

By careful and assiduous observation of the light changes shown by a variable star it is possible to compile what is known as a "light curve." This is simply a graph in which magnitude (i.e., brightness) is plotted against time. It is largely as a result of light curves that variable stars came to be grouped in a number of fairly well-defined categories. During their light cycles certain changes also occur in the spectra of variable stars and these, to a very large extent, are characteristic of each particular class of variable. Thus the long period variables we have just mentioned comprise almost exclusively types M, N, and S, whereas the Cepheids, named after the notable prototype δ Cephei show the spectra of types B, A, F, and G. This information tells astronomers that the former are "cool," red giant stars of considerable age and the latter, hot, white, or yellow young ones.

Variable stars can be conveniently placed in three major groupings as follow:

  a. Extrinsic variables
  b. Intrinsic variables, and
  c. Eruptive variables

Extrinsic variables, as their title might suggest, are *not* in fact true variables. Their apparent variability is due wholly to the fact that they are actually binary or multiple systems, the light fluctuations being a consequence of individual components in the same plane as ourselves eclipsing one another. Thus if a less bright component eclipses a more bright one, the total brightness of the system appears to diminish. The best known of these is probably Algol (the Demon Star) in the constellation Perseus, one of the prominent and well known north circumpolar constellations. Such variability, however, is in no way due to any inherent fluctuation in the light and energy output of the star. We will not therefore consider this category here since it would be entirely irrelevant. The intrinsic and eruptive variables being true variables (as their titles convey) are the categories in which we are presently interested.

Intrinsic variables are generally speaking single stars and, it is believed owe their light variations to changes taking place, for the most part, in the star's outermost layers. This major grouping can be subdivided into a number of distinct classes, the periods of which range from less than a day in the case of the RR Lyrae stars (named after the prototype RR Lyrae) to nearly five hundred days in the case of long period variables. These, of course, are average values. Brightness does not, on the whole, fluctuate vary widely except in the case of long period variables. Here again we are quoting average values.

Table 1 (Intrinsic Variables)

Type	Based on (Stars)	Average Magnitude Range	Period Range (days)	Spectrum
RR Lyrae	10	1.1	0.3-0.7	B8-F8
Cepheid	15	0.91	2.0-22.0	F0-R2
B Canis Majoris	7	0.08	0.16-5.40	B1-A7
Delta Scuti	5	0.17	0.07-0.19	F1-F6
Long Period	25	6.34	90.0-49.0	M, S
Semi Regular	20	1.77	—	M, N
R V Tauri	25	1.88	45.0-1244.0	F-M
Irregular	25	1.09	—	K-S

Eruptive variables can also be divided into a number of distinct classes. The fluctuations shown by these stars are generally greater, and indeed in many instances quite spectacular.

Table 2 (Eruptive Variables)

Type	Based on (stars)	Average Magnitude Range	Period Range (days)
U Geminorum	15	4.16	16-1,000
Z Camelopardalis	15	2.82	12-40
RW Aurigae	14	2.51	?
T Orionis	20	2.43	?
T Tauri	20	1.75	?
R Coronae Borealis	25	4.09	?
Flare Stars	23	1.39	?

Before we can dwell even briefly on the whys and wherefores of stellar variability it is necessary to say a little on the subject of stellar evolution. In this we are not really digressing from our main theme since planetary evolution must inevitbly be bound up with stellar evolution.

A star, any star, shows two major observable characteristics. These are brightness (generally termed magnitude) and surface temperature. When the former is plotted against the latter for a representative selection of stars we secure what has come to be known to astronomers as a Hertzsprung–Russell diagram. A typical example is shown in Figure 13, which also includes the

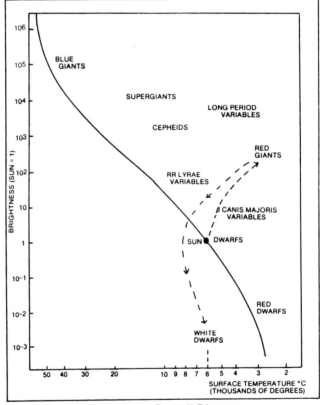

Hertzsprung-Russell Diagram

Figure 13

position of a number of the intrinsic variable classes. The prominent continuous line on the diagram is termed the "main sequence" so that those stars which by virtue of magnitude and surface temperature lie on or in close proximity to it are known, not surprisingly perhaps, as "main sequence stars." It should be noted at this point that our own familiar Sun is merely a rather average main sequence star with a surface temperature of about 6,000°C. Main sequence stars contain different amounts of material, though the *composition* of this material is the same for all. The position of a particular star on the main sequence is thus determined by its mass. It will remain on the main sequence only for as long as the nuclear reactions taking place within it provide the essential energy to keep it shining. After that things automatically begin to happen.

The fundamental process which provides the initial energy of a star after it has condensed from the primordial cloud of dust and hydrogen gas is known as the "proton chain" reaction. A condensing "embryo" star is composed almost exclusively of hydrogen. The hydrogen atom is the simplest of all the atoms, comprising as it does a nucleus consisting of a single proton and a solitary "orbital" electron. In our context it is the single proton nucleus with which we are really concerned. At the relatively low temperatures and densities prevailing at this stage of stellar evolution, a proton combines with another proton to form a deuteron, i.e., a nucleus comprising a proton and a neutron. At this juncture a positive electron (positron) is ejected. Another proton now links up with the deuteron to form a *helium* isotope. This, in turn, links up with another helium isotope to constitute a nucleus of ordinary helium. Coincidentally two protons are ejected to continue the chain reaction. The overall process is therefore the continuing transition of hydrogen to helium and the steady emission of energy. This fusion process can continue unabated for billions of years and represents the energy source by means of which the stars shine. It is also, incidentally, the process which briefly takes place in the thermonuclear or hydrogen bomb.

After a period, as central temperature and density continue to rise, a point is reached when the development of the star is stabilized. The star then attains the main sequence where it remains for the greater part of its active life. Chemically the star is still homogeneous. With the passing of time, however, as more and more of the hydrogen in the core is converted to helium, the temperature slowly rises to compensate for the depletion of fuel.

When nuclear fuel within the stellar core is finally exhausted, hydrogen transition must cease and the core begins to cool. Before long the crushing, inward force of gravity on the cooling core exceeds the outward thrust of the hot core gases. As a consequence the core begins to contract. By now it is almost completely composed of pure helium as a result of hydrogen "burning" (i.e., hydrogen to helium transition) which has been proceeding apace for billions of years during the main sequence stage.

During contraction the stellar core surrenders gravitational energy. This must go somewhere and the nearest convenient place for it is the envelope or outer concentric "shells" of the star. The gaseous envelope responds to this sudden acquisition of energy by colossal expansion. It is a fundamental point in physics that when a gas begins to expand it also grows cooler. Indeed the principle is used to practical advantage in industry for the liquifaction of gases.

Since stellar color is largely a measure of surface temperature we should hardly be surprised that a color change accompanies expansion. When white hot metal begins to cool it becomes, in turn, yellow, orange, and finally red. A cooling, expanding star exhibits a roughly analogous sequence, i.e., as the star swells it becomes red. In addition the vast increase in surface area of the expanding star means that less heat is emitted from each unit of surface area. This also constitutes a cooling process and further promotes the color change to red. Ultimately the star, by now well off the main sequence, has become, depending on its size, an orange or red giant.

Stars which have reached this grotesque, bloated stage have envelopes ranging from a hundred to a thousand times their former (main sequence) dimensions and surface temperatures at least 50 percent cooler. Our Sun will almost certainly emulate this performance some 5,000 million years from now. It isn't going to be pleasant but at least we won't be around to experience it. In so doing it will vaporize Mercury, Venus, Earth, and Mars and incorporate their gasified substance into its own. From the Sun or proto-Sun these planets came. To it they will in the end return. Jupiter should escape, but with such a roasting that it must be dramatically changed. By then our once yellow, quiescent Sun will have become a great, bloated red sphere of gargantuan proportions with a radius extending from its center out to the orbit of Mars or beyond. Such stars are already quite common in our skies, e.g., Betelgeuse (Alpha Orionis), Antares (Alpha Scorpii), Aldebaran (Alpha Tauri).

The helium core of a red giant continues to contract, and in time temperature and pressure become so high that helium nuclei fuse together to form carbon nuclei. Since this once again represents a loss of mass an emission of energy ensues. What we are now seeing is an analogous helium to carbon transition, a process generally referred to as "helium burning." Energy emission as a consequence of "helium burning" is however, much less than in "hydrogen burning."

The "ignition" of the helium core is a very important event in the life sequence of a star. When it occurs the temperature of the stellar core is uniform and a super chain reaction is initiated. In virtually a matter of moments all the helium nuclei are changed into carbon nuclei. This event is known as the "helium flash," but since it takes place within the core of the star it is, not surprisingly, invisible. It is something which no observer can ever hope to witness although we have no reason to doubt that it really happens. Following the helium flash the red giant derives its energy from hydrogen and helium "burning" in separate concentric shells surrounding the core.

The precise evolution of a star beyond the helium flash stage remains uncertain. It would appear to move to the left and downward on the Hertzsprung-Russell diagram, briefly recrossing the main sequence as it does so. We can be fairly certain, however, that it ends up as a white dwarf.

Energy output of white dwarf stars is so low that they continue to shine by radiating away their final reserves of energy. In the end there is none left to emit. A star's final stage, its virtual death, is probably that of a sphere not much larger than the Earth but having a density estimated at fifty thousand times that of water. It might just be covered with a thin layer of ice and surrounded by an atmosphere of hydrogen and helium a few feet thick.

The path followed by a normal star (in this case the Sun) after it quits the main sequence is shown by the dotted line on Figure 13. From this diagram the position of certain of the groups in the intrinsic variable category will be seen to appear, i.e., RR Lyrae, B Canis Majoris, and Cepheids. It will be apparent, therefore, that these stars lie outside the normal gamut of stellar evolution. In view of their erratic and unusual behavior we may hardly be surprised at this. It is believed that the intrinsic variables owe their variability to pulsation. Some of these stars show spectroscopic emission lines of hydrogen, which could be taken for evidence of the existence of a shock wave in the upper layers or shells of the stellar "atmosphere." Such emission lines could conceivably owe their existence to hot gases moving outward from the star's surface behind the wave front. There seems little doubt that the Cepheids are pulsating variables. The light cycle of the great majority of these stars is so constant that it can repeat itself to within a fraction of a second!

Some of the RR Lyrae type variables are also extremely regular in their fluctuations. It was once thought that the entire class showed this characteristic, but, today, with the advent of more sophisticated techniques for the measurement of stellar magnitudes, it has become increasingly apparent that only a small proportion of RR Lyrae stars can be regarded as having constant periods and amplitudes.

Both Cepheids and RR Lyrae variables show change not only in brightness but also in spectral type, temperature, size, and color during the course of their light cycles. Considering once again the Hertzsprung-Russell diagram and its axes showing magnitude (brightness) against temperature, or color, we see that such stars are not sticking to one point on the diagram, as for example the Sun does, but are in fact oscillating slightly with respect to both axes of the diagram.

As a general rule maximum brightness corresponds with maximum velocity of approach (measured by Doppler shift in the spectrum lines). The shock wave phenomena mentioned a few paragraphs back must, however, also be taken into account. The variation in the size of the star which accounts for the change in brightness is not considered to be large and probably only amounts to about a tenth of the star's diameter. In fact it is perfectly feasible that only the outer layers of the stellar atmosphere are involved in these volume changes. In other words the *entire* star may not be expanding and contracting.

We saw from Figure 13 that Cepheids and RR Lyrae variables lie at points on the Hertzsprung-Russell diagram which indicate that their evolutionary pattern differs in some degree from that of conventional stars. The very fact that they show variability is in itself a manifestation of this fact. It is reasonable then to ask at this point just *why* these stars should be so different, *why* they should be variable. It must be conceded, though, that at present we are still largely in ignorance of the causes for these fluctuations.

Let us return at this juncture to the early stages in the evolution of a star. A transition from hydrogen to helium begins. The amount of hydrogen decreases with time while that of helium increases in direct proportion. There was a period when it was thought that interior convection currents slowly circulating effected a thorough mixing of hydrogen and helium. Today this concept is discounted, though magnetic fields within a star may permit a modicum of hydrogen-helium mixing.

It can be shown theoretically that a complete mixing of hydrogen and helium within a star would cause that star to evolve differently. This difference would manifest itself on the Hertzsprung-Russell diagram as a slow move to the *left* of the main sequence (which in effect implies an *increase* in surface temperature). *Lack* of hydrogen/helium (and this is much the more probable) causes a star to move along a line to the *right* of the main sequence (indicating a *decrease* in the surface temperature (Figure 13). During the early stages of stellar evolution helium is produced more rapidly at the star's center, i.e., at the core. This is hardly surprising in view of the fact that it is there that temperature and density must be at their greatest. In the fullness of time a core of pure helium is formed and energy ceases to be evolved in this region of the star. At this point a "skin" forms around the core which is then compressed. As a consequence its temperature is further increased, which in turn initiates more nuclear reactions within the skin ("skin" is probably a less than perfect term yet it is one which, in the circumstances, is peculiarly apt).

In time this compression or shrinkage of the core comes to a halt. Were it to proceed, pressure in the core region of the star would become so great that it would overcome the weight of the overlying shells of gas. In other words equilibrium is reached. This equilibrium is highly critical and depends on the fact that the mass of the star's helium core must *not* exceed 1.44 times the mass of the Sun. This optimum figure is known as Chandrasekhar's Limit after its discover. Should this limit be exceeded the result is a supernova. Little imagination is required to assess the effect of this on a planetary retinue. Were it by some inconceivable misfortune to occur to our Sun all the planets would be vaporized in a matter of hours. There would be no exceptions. Not even Pluto would escape. Lest this baleful thought cause alarm and despondency among readers we would hastily add that there is very little likelihood, virtually none in fact, of the Sun's ever becoming so unkindly disposed toward us.

From Figure 13 it is apparent that R R Lyrae variables occupy a specific region in the evolutionary curve. Stars like the Sun eventually leave the main sequence, proceed to the red giant region, than pass back over the main sequence to wind up in the white dwarf areas. Simplifying and abbreviating what is undoubtedly a long and involved story, the path followed by certain stars when they leave the main sequence take them into one or other of the "variable" regions (i.e., B Canis Majoris, Cepheid, or R R Lyrae). These are regions which stars like the Sun effectively miss. Such forms of stellar instability are thought to affect stars which have already passed through the red giant state. The upheavals are probably due to internal readjustments between thermonuclear reaction and thermonuclear fuel. In fact they have been likened to flare-ups in half-burned logs or pieces of coal in a domestic fire. This homely analogy seems quite an apt one. It is a fairly well-established fact that the R R Lyrae stars, for example, show characteristic oscillations, i.e., variability manifestations, yet stars on either side of this critical point on the curve (Figure 14) do not. This tendency to display differing degrees and forms of variability during certain periods in their evolution is, as we have already mentioned, something which we do not wholly understand at the present time. *What* is happening is obvious enough—brightness fluctuations by the star, probably in its outermost layers. What is still far from clear is *why* these

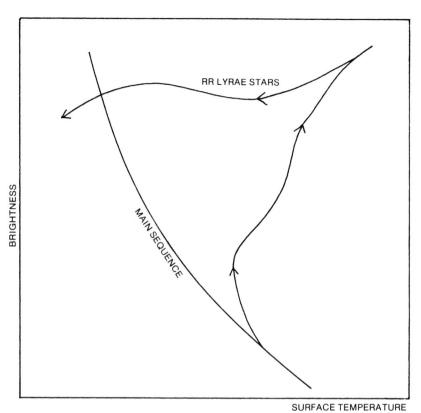

BRIGHTNESS

SURFACE TEMPERATURE

RR LYRAE STARS

MAIN SEQUENCE

Figure 14

fluctuations take place. We see the effect but would like to know the cause.

This generalization is as true when applied to the eruptive variables. This category however is prone to much more violent and cataclysmic outbursts. The U Geminorum groups (named after the prototype U Geminorum) are particularly renowned in this respect and have for this very reason been termed "dwarf novae." One of the best known in this group is S S Gygni which can rise from a minimum of 12.1 to a maximum of 8.1—four clear magnitudes. All stars in this class are noted for their violent fluctuations and their irregular timing. S S Cygni, for example, averages 50.4 days between maxima giving six or seven maxima per year. During 1960 however *twelve* maxima took place, which meant the mean period for that particular year falling to only about thirty days—an uncertain sun indeed! A graph of the star's fluctuations (i.e., its light curve) taken from my own observatory records for the period August 1964 to November 1964 is shown in

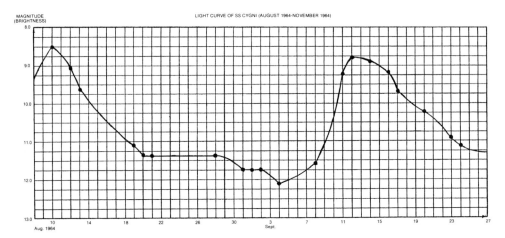

Figure 15a

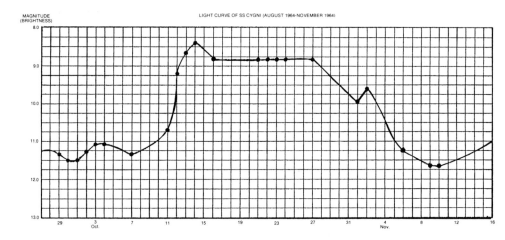

Figure 15b

Figure 15. A maximum of 9.5 took place around August 10. This was followed thirty-two days later by a further maximum of 9.25 on September 12. The next maximum took place on October 14, again after a lapse of thirty-two days. This time maximum was 9.6.

On this occasion the highly unpredictable characteristics of these stars was shown for, though by October 16 brightness had begun to wane, the star remained on a "high plateau" of 9.2 for eleven days. When the drop to minimum got under way again around October 27 this was erratic due to a sudden brief rise in brilliancy of approximately 0.4 between October 31 and November 1. This star, and those like it, are among the most fascinating to the regular variable star observer. Within certain limits virtually anything can happen, and frequently does. With the more regular variables an observer can often anticipate what he is going to find when he goes to the telescope, but the totally unpredictable variations in magnitude and length of period make the U Geminorum stars something rather special. This may well account for their popularity among the ranks of the world's variable star observers, that small, silent, dedicated band who, under the great dome of night, cheerfully pursue their hobby while the rest of the world, unheeding and uncaring, sleeps.

Among the eruptive variables there are of course variations just as there are among the intrinsic variety. In the intrinsic category (Table 1) we find there are differences in behavior between the Cepheids, R R Lyraes, and B Canis Majoris, to name but three. Similarly among the eruptives, U Geminorum stars differ substantially from the Z Camelopardalis type which, in turn, differ from the T Tauri type (Table 2). In a book of this nature it would be as impossible as it would be undesirable to cover each category in detail. A whole book could be written on the subject of variable stars and even then justice would have by no means been done to the subject. It is a field in which much remains to be done and even more to be discovered. What we have tried to do in the light of a still imperfect understanding of the nature of variability is to give a brief impression of its forms, its place in stellar evolution and its implications—most important, so far as these pages are concerned, are the implications in respect of *planets*.

There seems no valid reasons why at least some variable stars should not have planets. Planets, as we have already seen, are probably a natural and perfectly normal adjunct to stellar formation, either coincidentally with the star from the same primordial gas cloud or spawned from the star itself in the course of evolution. Stars, which later in their lifetimes become variable, are in all probability just as likely to have acquired planetary retinues. Later on they become very cruel parents indeed. It might be argued in this respect that in blasting Mercury with perpetual withering heat our Sun is hardly being particularly merciful to its innermost planet. Nor are its feeble rays bringing much warmth or light to the surface of poor, lonely Pluto. But neither is it alternately roasting *and* freezing these planets whilst it is undoubtedly being very benign to Earth. A *variable* Sun could hardly thus be described in respect of Earth.

It has often been asked whether there are any circumstances in which a planet might enjoy reasonable surface conditions should it happen to orbit a variable star. There seems but a single possibility here. Let us assume that a certain star is variable and that its fluctuations are not too extreme. This tends to rule out any of the eruptive category. The highly idealized circumstances are portrayed in Figure 16. A planet orbits this star, its orbit,

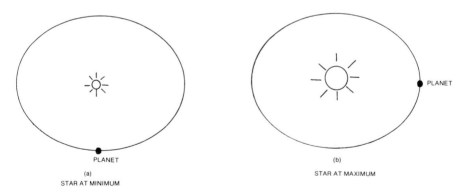

Figure 16

however, being highly elliptical. Suppose then that when the star is at minimum the planet is as shown in (a) i.e., at perihelion, at nearest point to the star. By the time the star has gone to maximum as in (b) the planet is at aphelion and furthest from the star.

Clearly there must be many ifs and buts relating to a situation such as this. First of all it would be absolutely essential that the status quo be maintained, i.e., planet always furthest from star at stellar maximum, nearest to star at stellar minimum. This seems just too perfect to be possible.

Suppose, as seems very likely, these two factors began ever so slightly to slip out of phase. With the passage of time the planet would no longer be at its maximum distance from the star when the latter was at maximum brightness. The ultimate effect could be shown in Figure 17 in which the planet is *closest* to its sun at

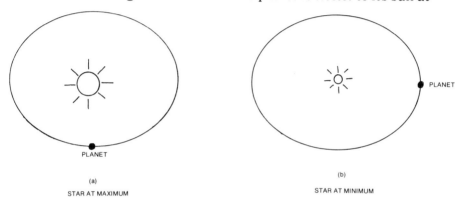

Figure 17

maximum and *furthest* from it at minimum—the exact reversal from the happy position portrayed in Figure 16. Surface conditions on the planet could then be intolerable, a period of withering heat followed by one of relative chill. For the conditions shown in Figure 16 to apply, the star's period of variability would have to be very regular and be also of such duration as to coincide exactly with the planet's orbital movements. There may be instances in this and other galaxies where these conditions are satisfied but they must be very few and far between. They would virtually have to be experienced to be believed—and it seems a shade unlikely that any of us now living will be permitted to secure this sort of proof!

Clearly the distance separating a planet from its central luminary, vitally important even in the case of a non-variable star, is going to be considerably more so in respect of a variable star. A planet orbiting a variable star akin to the Sun in temperature and size and at the same distance from it as Earth is from the Sun is going to experience alternating periods of withering heat and considerable chill. If the planet should lie closer to the star the chill period might prove temperate and the hot period (maximum of the star) fierce beyond belief. A more remote planet, at a distance comparable to Mars in our own system, might find the period of minimum brightness one of frigid cold and of maximum brightness as pleasantly temperate. To planets of such a star as remote as Pluto or Neptune it is doubtful if there would be a great deal of difference other than what is normally just an exceptionally bright star in their skies becoming periodically brighter. To most of the intrinsically variable stars of the types we have mentioned, distance separating luminary and planets must be the dominating factor. To this we must of course add eccentricity of orbit. As we saw in the idealized case just discussed a highly elliptical orbit means that distance between star and planet will also be variable. This parameter could prove even more variable were the star not at the center of the ellipse. An extreme example of this kind of thing within our own system (apart from comets) is seen in the asteroids of the Apollo family. The fairly recently discovered Icarus is a classic example which at aphelion is 183 million miles distant from the Sun and at perihelion only 18 million miles and almost certainly by then glowing red hot. The possibilities are outlined in Figure 18.

As we saw from Table 1, with the exception of the long period variables, most of the other types of intrinsic variable have average magnitude ranges which do not represent tremendous fluctuations. For example an average of seven typical β Canis Majoris stars show an average fluctuation of only 0.08 δ Ceti, one of the seven, has indeed an average magnitude fluctuation of only 0.03 though this rises to 0.17 in the case of B W Vulpecula. Planets orbiting such stars at a reasonable distance from them, though experiencing temperature fluctuations, might not find these too extreme. In the case of the β Canis Majoris type periods range (on the basis of the same seven stars) from 0.16 days to 5.40 days. Fluctuations would thus be frequent. β Canis Majoris stars are however young, fairly hot stars (spectral class B1–A7) and on the basis of the presently accepted theories of planetary formation

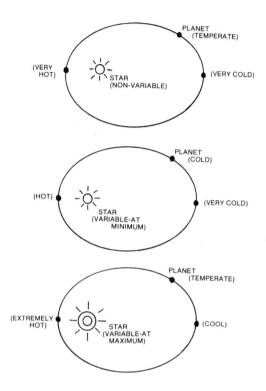

Figure 18

these are not stars which we would expect to possess planets. This is true also in respect to the R R Lyraes and probably the δ Scutis. The other instrinsic variables shown in Table 2 could, however, qualify.

When we come to the eruptive variables we must, almost of necessity, envisage the most fiercesome conditions on any planets that happen to orbit stars in this category. In this case average magnitude ranges are very much greater. Fifteen stars of the U Geminorum class show an average magnitude range of 4.16. This is considerable. Small wonder that these stars tend to be regarded as "dwarf novae." Of the fifteen selected stars, the prototype U Geminorum leads the field at 5.6 followed closely by

S W Ursa Majoris at 5.4. Lowest of the fifteen are X Leonis and R U Pegasi at 3.1. If planets exist around any of these stars then, unless they lie very far out, their surfaces can only be a wilderness of scorched cinder and heat blasted rock—desolation beyond belief. The notorious Death Valley in California might well appear as a lush paradise in comparison. In some cases rocks could become red hot. We can envisage glowing cliff faces and perhaps pools of molten lead, zinc, and tin. Even trickles of liquid gold (melting point 1,063°C) are feasible. We doubt however, if such an environment would lead to an interstellar gold rush. Panning among the cool streams of Earth for the yellow metal seems infinitely preferable.

The possibility also of nearly incessant, probably cataclysmic seismic and volcanic action cannot be ignored. As the star approaches maximum and the outer incandescent shell pulses closer to the planet, considerable strain must be imposed on the planet's tortured crust. The degree of intensity would in all probability be directly propositional to the extent of the particular maximum. The surface of such a world could be pockmarked by large and small volcanic cones and lacerated by a contorted network of fissures and rifts.

The alternating effect of tremendous heat during maxima and a certain degree of cooling during minima could very easily lead to large scale thermal disintegration of the rocks. Much would depend on the materials of rock formation, some rocks lending themselves more easily to this type of effect than others. Nevertheless large scale disintegration of this nature seems a very likely possibility and one which could, in a geologically short space of time, wear down great mountains and lay bare much of a planet's mineral wealth.

It is hardly likely that atmospheres, irrespective of their constitution, could survive on such planets. Moreover the withering heat and intense light deluging the surface would not, by any means, represent the sum total of received radiation. We must also take into account the possibility of a flood of gamma, ultraviolet and X-rays.

It must be clear that the chances of life forms of any kind existing on planets orbiting eruptive variable stars are zero. The circumstances would not allow of its initiation and even if, by some miracle it did start, its development would be surely cut very short. Indeed it is highly doubtful whether temporary visitors, astronauts representing a highly advanced technological society, could survive, even briefly, out on the surfaces of such cruelly tortured worlds.

In painting this devastating picture of probable conditions on worlds orbiting eruptive variable stars we must avoid giving the impression, however unintentionally, of implying that planets of the *intrinsic* variables are going to be all that pleasant. Many of the effects we have just enumerated could also happen there, albeit perhaps to a somewhat less degree. It is possible to postulate planets which, when at maximum distance from their sun or when that sun is at minimum, have atmospheres, perhaps even seas of a sort. Great turbulence in that atmosphere due to incessant heating and cooling seems a near certainty. And what of seas and oceans were they to exist? Atmospheric turbulence producing winds of frightful intensity or earthquakes on the ocean floor could throw colossal tsunami-like waves inland. Increasing heat as star rose to maximum would certainly result in large-scale evaporation of any seas, rivers and lakes the planet possessed. Thus we could reasonably anticipate evaporated lake and sea beds rich in deposits of crystallized magnesium, potassium, and sodium salts. All this evaporated water in the now steamy, highly humid atmosphere would likely condense in a tremendous torrent when star maximum was past and cooler conditions were once again sweeping over the planetary surface.

Biological potential might exist on certain planets of intrinsic variable stars assuming that the stars' fluctuations were of a sufficiently low order. In this respect we are thinking also in terms of stars having regular periods, the Cepheid variables for example, might suffice. Whether amid these relatively less extreme conditions life could obtain a hold and develop must be a matter of considerable conjecture. If at some point in the galaxy this has happened it might be more easy to envisage a form of troglodyte society, either a low order one dwelling and sheltering in caves or crude subterranean passages or, if by chance such a society had achieved a measure of advanced technology, then a system of hollowed out underground cavern communities with air shafts to the surface, interconnecting tunnels, perhaps even with electrically powered vehicles running along them.

The fact remains, nevertheless, that it is probably unwise to countenance or to speculate on life under such conditions. A day could come however, when some of our descendants eventually reach such worlds. Nor can we entirely ignore the possibility that other galactic communities, greatly in advance of our own, may already have done so. In this respect life on such planets (more correctly, brief sojourns there for purposes of scientific investigation) must be considered.

(9)

Cosmic Wanderers

OUR AIM IN this chapter will be to discuss two possibilities, one natural, the other artificial—stray planets, and artificial planetoids respectively. It is as well to state at the very outset of course, that, so far at least, we have no actual experience of either. Consequently we see them only for the moment as possibilities.

And so to the first of these, stray planets. A good plan in the first instance might be to dip into the pages of science fiction. A favorite theme of "sci-fi" works over the years has been Doomsday, the end of the world, the destruction of our Earth by one means or another. An obvious front-runner in these stakes has been the theme of a planetary collision. To destroy Earth utterly by such means entails collision with our world by a body of fairly reasonable mass, by something say of the proportions of the larger asteroids at least, probably even of the Moon, Mercury, or Mars. Unfortunately (though extremely fortunately for us) such bodies do not take to jaywalking about the Solar System. The only exceptions are the so-called Apollo group of asteroids, the members of which from time to time cut across the orbits of several planets, including that of our own. In so doing they constitute a potential menace though mathematically the odds are still heavily weighted in favor of mankind. Of these interesting bodies and their peculiar orbits more anon.

Since collision with a wandering Apollo asteroid might devastate an entire continent on impact, or raise vast tsunamis by plunging into an ocean, the theme is, from the viewpoint of the "sci-fi" writer, one of considerable interest. It is most unlikely however that such a dire event would totally destroy our planet. In these circumstances fiction writers must cast their nets wider. What could be more convenient than the intrusion into the Solar System of a large planet which has somehow, aeons ago, managed to become detached from the planetary system and gravitational attraction of another star? After an eternity of aimless wandering in the intense cold and blackness of interstellar space this cosmic bull-dozer approaches the Sun and is, as a consequence, "captured" by the gravitational pull of our star. And so, of course, we have the holocaust to end all holocausts when the intruder either blunders into Earth or totally devastates our planet's surface by virtue of a near miss before either falling into the Sun or somehow escaping back into the stygian interstellar realms from whence it came.

This is a theme which has caught the imagination of fiction writers (not to mention fiction readers) almost from the beginning. Two classic examples of it stand out and are certainly well worth mention. Pride of place should probably go to the late H. G. Wells whose short story entitled "The Star" written in the closing years of the last century still exercises a very considerable appeal. Despite the title, the body concerned is no star but a planet of fairly considerable mass.

The planet is detected by observatories in several continents during the first days of the year. Interest deepens and excitement quickens when it becomes obvious that the intruder is about to collide fairly and squarely with the huge outer planet Neptune. This it does and Wells dwells briefly on the activity within observatories, the rush to expose photographic plates and gather spectroscopic apparatus to record the novel and astonishing sight of a world's destruction. Writes Wells, "for it was a world, a sister planet of our Earth, far greater than our Earth indeed, that had so suddenly flashed into flaming death." The heat of the concussion, it appears, has transformed the two solid bodies into one vast mass of incandescence. This then is the "star" which now starts its final plunge towards the Sun. So far as Earth is concerned there is no cause for alarm for our world is not in the way—or so it seems! It is at this point that we make our first acquaintance with the only character in the story, described by Wells as "the master mathematician." This nebulous yet strangely real person completes calculations which prove incontrovertibly that the gravitational influence of "the mighty planet Jupiter and his moons sweeping splendid round the Sun" will divert the glowing intruder and the fused mass of its first victim toward Earth. Nemesis for Earth is at hand! In the words of Wells "The streets and houses were alight in all the cities, the shipyards glared, and whatever roads led to high country were lit and crowded all night long. And in all the seas about the civilized lands, ships with throbbing engines, and ships with bellying sails, crowded with men and living creatures, were standing out to ocean and the north. For already the warning of the master mathematician had been telegraphed all over the world and translated into a hundred tongues."

The story, being in the nature of a scenario, permits vivid word pictures of the dire event to be painted. Gradually the climate becomes hotter as the intruder grows nightly in our skies. Upon all the mountains of Earth snow and ice begin to melt. Rivers become swollen and turbid, bearing swirling trees, débris and soon the bodies of men and of beasts. Tremendous tides and storms, unprecedented in human history, lash coasts and drive inland, drowning whole cities. Hillsides slide, fissures open, volcanoes erupt cataclysmically. The whole side of Cotopaxi in the Andes slips out in one vast convulsion allowing a tumult of lava to pour out. All seems lost. It is the end. Then with the final cataclysm imminent, the Moon comes between Earth and intruder. At that point intruder and Earth are at their nearest and swing about one another. Then the danger is passed and now the strange intruder proceeds on the final stage of its headlong journey downward into the Sun. Earth is saved but it is admittedly a rather different Earth that emerges. Days are now hotter and areas such as Greenland are covered in lush vegetation. The Sun also is larger and the Moon, now shrunk to a third of its former size, takes eighty days between new and new.

It was not until the year 1932 that the theme of planetary collision was really well tackled again. It was then that Philip

Wylie and Edwin Balmer wrote their epic science-fiction novel, *When Worlds Collide* (produced also as a very thrilling and entertaining movie in the early fifties). This is a much longer and more detailed narrative than H. G. Wells's "Star" and is built around a number of real characters. There is even a love interest running throughout the story in the shape of the attractive Eve Hendron.

The intruders (we have to use the plural in this instance) are a pair of stray planets revolving around a common center of gravity in a fashion similar to a binary star system. These approach, in the best science fiction tradition, from the depths of interstellar space. One is very large, somewhat like Jupiter, the other smaller and much more Earth-like. The larger of the pair is named Bronson Alpha and the smaller, Bronson Beta after their discoverer, Dr. Sven Bronson. Since they are first detected in the skies of the southern hemisphere where (in 1932 at least) there are few large observatories, some little time passes before their true nature and the dire peril they represent is fully appreciated. Eventually with calculations complete, checked, double checked and checked again (this was also before the days of computers) a number of highly unpleasant facts become unmistakeably clear. The Bronson bodies, as they have come to be termed, will, in the first instance, make only a very close approach to Earth. This will have the effect of raising huge tides in the oceans as well as causing earthquakes and volcanic eruptions on the grand scale. They will then orbit the Sun and head back towards Earth. On this second occasion there will be no escape for our world. The larger of the pair, Bronson Alpha, will collide fairly and squarely with Earth.

And this is indeed what happens. On the occasion of the first passing the geography of Earth is changed radically and the convulsions which shake the planet are as tremendous as they are devastating. Untold millions perish and the Moon is destroyed. But already on a geologically stable portion of the United States (in Michigan in fact) Cole Hendron and his team are fast completing a space-ship (named the "Ark") which, it is hoped, will be able to carry a limited number of young men and women plus representatives of terrestrial flora and fauna to Bronson Beta, the smaller, Earth-like intruder. According to all calculations Bronson Alpha will, after destroying Earth, return to interstellar space leaving Bronson Beta to orbit the Sun more or less as

Earth once did. The celestial mechanics involved in all this might not today stand up well to close scrutiny. Indeed the whole project sounds extremely risky and those contemplating transfer from a doomed planet to a by no means safe alternative seem, throughout the proceedings, to be incredibly optimistic. However this may be understandable since it represents the only possible hope of salvation. Drowning men tend to clutch at any straws.

In the event all goes according to plan. Earth is destroyed as forecast but one of the two space ships from the United States reaches Bronson Beta safely. The fate of its sister ship plus those from other nations is not disclosed but one is left with the impression they didn't make it. The description of the actual collision between Earth and Bronson Alpha is very graphic. In the final moments our planet bulges and becomes plastic. Great cracks rend its surface, then a huge chunk tears loose. Within seconds the actual impact takes place. Most of Earth is fused and absorbed by Bronson Alpha, the remainder becoming a number of small angular asteroids destined to orbit the Sun in individual orbits.

Both these stories envisage bodies from interstellar space entering our Solar System. This can only be the result of planets somehow escaping from the gravitational attraction of their parent stars and setting off on a very long and cold journey through interstellar space, a journey which, considering the distance separating stars, could quite easily go on forever. The chances of an errant planet coming within the gravitational attraction of another star are slim indeed. In fact a graphic illustration of the position can be given by illustrating what is considered the likely effect of two *galaxies* colliding. On the fact of it this seems like a most awesome cosmic event, conjuring up visions of millions of stars all colliding with each other, but in fact the truth of the matter is quite the opposite. So widely spaced are the stars in galaxies that it is confidently believed that one galaxy could pass safely through another with few if any stellar collisions taking place.

We can see fairly plainly therefore how remote are the chances of errant, wandering planets coming under the influence of other stars. We cannot claim it is impossible. We must assume there is always the one in several million chance that such an eventuality could take place. In the realms of science-fiction it is easy to arrange for this, but only there. Thus, even if it is possible for certain stars for one reason or another to shed one or more of their planetary retinue, it still seems highly improbable that the resultant "rogue" planets would ever constitute any danger to ourselves or to the beings in any other solar system. Later in the chapter we will consider the possibilities of planets escaping from the gravitational clutches of their parent suns. Meantime we will give a little thought to an item mentioned earlier—"rogue" planets or more correctly "rogue" asteroids loose within the confines of our own system. We refer of course, to the so-called Apollo series of asteroids many of which intersect the orbit of Earth. Some thirty of them are now known and though we have referred to them as asteroids, asteroid-like might be a better definition since they may simply be the nuclei of comets that have lost their volatile components.

During the closing days of October 1937, a body discovered by Dr. Karl Reinmuth of Heidelberg University and therefore tentatively designated "Object Reinmuth 1937," passed within 800,000 kilometers (500,000 miles) of the Earth. This is roughly twice the distance between Earth and Moon, and though it may seem like an immense distance it is no such thing astronomically speaking. News of the event only reached the news media in the early days of January 1938, when it brought forth such newspaper headlines as "Earth narrowly escapes collision with another world." Despite the flambuoyance of such comments they were not too far wide of the mark. Had such a collision taken place, the energy released would, it is reckoned, have been equivalent to the simultaneous detonation of ten thousand 10-megaton hydrogen bombs and would have created a crater some 20 kilometers (12.5 miles in diameter). It is a fact however that such catastrophies are so infrequent that none has been recorded within the period embraced by human history. Object Reinmuth 1937, subsequently named Hermes, has an estimated diameter of one kilometer (about 0.6 mile).

Hermes was not the first asteroid of its class (they are frequently referred to as "Earth grazers"). The first such object had been discovered five years earlier, also by Dr. Reinmuth, who was at that time carrying out a photographic search for ordinary asteroids, i.e., small planetoids orbiting the Sun between the orbits of Mars and Jupiter. This object was provisionally designated 1932 HA, but later named Apollo after the Greek solar deity. Thus the term "Apollo" was subsequently given to the entire class of bodies having Earth-intersecting orbits.

It seems 1932 was a good year for this sort of thing for another small asteroid, later named Amor, was also discovered. It has an orbit similar to that of Apollo except that its perihelion or nearest point to the Sun is 1.08 astronomical units from that body. Since the distance between Earth and Sun is one astronomical unit (A.U.) Amor will not cross the orbit of Earth. Or so at least it was thought. Subsequent investigation into the whole question revealed that orbits of objects having perihelions close to 1 A.U. often evolve into Earth-intersecting orbits.

Since 1932 a total of twenty-eight Apollo-like objects have been discovered and a rather smaller number of the Amor type. It is not always easy to tell whether or not an object belongs to the Apollo-Amor class since the orbits are not at all easy to determine. Even when they are, the object concerned tends to get "lost" and is only rediscovered at a much later date. For example, Apollo, formerly 1932 HA, was not relocated until 1973, over forty years later. The same goes for Adonis, the second such body to be discovered (in 1936), which was not traced again until 1976 by which time it must have crossed Earth's orbit more than thirty times. We must assume that on these occasions Earth was far distant from the intersection point.

At one time it was believed that Apollo-Amor objects were either bodies somehow ejected from the broad asteroid belt (2.1-3.5 A.U.s from the Sun) or bodies which, for reasons unknown, had been pursuing their peculiar orbits since the early aeons of the Solar System. The asteroid belt between the orbits of Mars and Jupiter is believed to include a total of 400,000 objects having a diameter in excess of one kilometer. It is indeed possible that some of these bodies could become Apollos. However a recent hypothesis suggests that many Apollo-Amor objects are not asteroids at all but the remnants of comets.

At present about four Apollo bodies per year are being discovered, so that the total number of these "rogue" planetoids is probably quite considerable. To have a diameter of one kilometer or more an Apollo must have a magnitude of at least +18. Several estimates of their number have been made but that due to Shoemaker (750 ± 300) is presently recognized as the most valid.

It is entirely a matter of chance whether or not an Apollo object will collide with the Earth. It can be said, however, that at the present time none of the known Apollo objects is on a collision course with our planet. However, it is an established fact that all the Apollo-Amor bodies are continuously influenced by the gravitational influence of the nearby planets, particularly that of the giant planet Jupiter. This has the effect of causing the asteroidal orbits to precess. The effect of precession is to cause the major axis of elliptical orbits to rotate gradually through 360°. Thus all objects having a perihelion *within* the orbit of Earth and an aphelion beyond it will ultimately pursue an orbit of Earth and an aphelion beyond it will ultimately pursue an orbit that intersects with that of Earth. On this basis on Apollo object should find itself in an orbit that intersects Earth's orbit about once every 5,000 years. This does not mean, of course, that Earth is going to be at that point at that time. After all it is perfectly safe to cross a railroad track if the train is not in sight and perhaps still fifty miles up the line! In most cases our planet will be at some other point in its orbit when the Apollo object is at the point of intersection. What then, it may be asked, are the chances of collision, of both Earth and asteroid being at the same point in both space and time? According to recent calculations, for any given Apollo object, collision probability works out at around once in 200 million years. Assuming that there are 750-1,000 Apollo asteroids larger than one kilometer in diameter, approximately four should strike the Earth per million years. These seem reasonably good odds for mankind especially if in the last million years we have already had our four asteroidal strikes.

What probably represents our most recent encounter with a small asteroid took place between 25,000 to 50,000 years ago in the American Southwest. The result was a crater slightly more than one kilometer in diameter and almost two hundred meters deep—the famous "Barringer Meteor" crater in Arizona. The small asteroid was probably about 100 meters in diameter. The youngest known crater larger than ten kilometers in diameter constitutes the basin now occupied by Lake Bosumtive in Ghana. This collision probably occurred about 1.3 million years ago.

The consequences of such collisions can only be described as dramatic. The kinetic energy of an Apollo object having a diameter of one kilometer and a density of 3.5 grams per cubic centimeter (a fairly average figure), is estimated at 4×10^{27} ergs. This is about 100,000 times the energy released by detonation of a one megaton thermonuclear warhead. Since on impact all this energy has to be dissipated in one way or another, the results must clearly be spectacular. Only a small fraction of this energy is required to achieve vaporization of the impacting body. Most of it will be expended in pulverizing the material of Earth's surface and ejecting the product outwards and upwards at a considerable velocity.

This being a planet having a thick atmosphere it is hardly surprising that the consequences of past asteroidal impacts have been largely worn away by millenia of erosion due to wind, rain, and ice. Nevertheless the remains of about fifty such impact structures have been located and identified on the surface of our planet.

The next question to arise concerns the origins of these peculiar bodies. At one time it was believed that they had existed since the formation of the Solar System 4.5 thousand million years ago. Had this been so, however, it is reasonable to assume that by now there would be few survivors among a family of objects that make a point of crossing the orbits of major planets. It might be argued, of course, that the present population of Apollos (some 700–800) represent merely the remainder. This idea is ruled out by virtue of the fact that cratering of the Moon's airless surface indicates a fairly continual flux of impacting objects and not a trend that has tailed off.

The inference therefore is that these Apollo bodies are somehow being injected into the *inner* Solar System from some source or other and at fairly constant rate. This rate, so far as bodies with a diameter greater than one kilometer are concerned, is put at around fifteen per million years. This, it is reckoned, should balance removals due to collisions, for during this period an estimated four asteroids will have collided with Earth, three with Venus, and one with the Moon, Mercury or Mars. This adds up to eight. What then of the remaining seven? It is believed that these are perturbed into hyperbolic orbits and, as a consequence, ejected completely from the Solar system. Thus we have a link between "stray" bodies *within* the system and *true* strays sent off into interstellar space. Such strays of course, would not represent the hazard to planets in neighboring solar systems as did those in the two classic science fiction works we described earlier. As Doomsday agents they are a bit on the small side, though we on Earth would certainly not like to be sideswiped by one of them.

It would appear that our estimated fifteen Apollo objects per million years injected into the inner Solar System are either small asteroids from the main asteroid belt between Mars and Jupiter which have somehow strayed, or they are the remnants of former comets. Both sources could be responsible for the supply.

Within the asteroid belt proper collisions are a fairly frequent occurrence. Of the 400,000 members larger than a kilometer it is estimated that about 400 are being created or destroyed by collisions every million years. It is considered unlikely however that these collisions would in themselves cause ejection from the asteroid belt into the inner Solar System. Some other much more potent agency is required. This could be provided by resonant interactions increasing the orbital eccentricity of these small bodies so that eventually they come under the gravitational influence of Mars which in turn supplies the necessary "push."

The "comet" theory could be even more valid. Though comets contain in their nuclei large volumes of solid carbon dioxide and ice, there is also a high proportion of rocky or metallic material present. On approaching the Sun carbon dioxide gas and water vapor are volatilized and streamed off into space along with a dust of solid fragments. On those occasions when this dust enters our atmosphere we are treated to the brilliant and beautiful spectacle of meteor showers. This dust, however, represents only a small fraction of the metallic or rocky material in the cometary "head" or nucleus. Should the nucleus split (and this has been observed on occasion) large chunks of solid matter are "unloaded" into space. Cometary heads only rarely split however, so that on each occasion when the comet nears the Sun and releases its carbon dioxide, water vapor, and dust, the head becomes proportionately richer in metallic and rocky matter.

Eventually there must come a stage when the comet has lost all its carbon dioxide and water. At that point is is no longer a comet but an asteroid of sorts. Obviously this will come to pass more quickly in the case of short period comets, i.e., comets which return to the environs of the sun fairly frequently. Short period comets, on the whole, do not have aphelion points (points furthest from the Sun) much beyond the orbit of Jupiter. This is interesting and rather suggestive since all the Apollo objects have similar aphelion points. Now it must be fairly obvious that mere depletion of volatiles from a comet, though it may in time transform it into an asteroid, will not give it the orbit of an Apollo object. But if short period comets already have the orbits of Apollo objects then only the volatilization of the carbon dioxide and water is necessary to render them true Apollo-type asteroids.

It is time now, however, to resume our original brief which is that of stray planets *beyond* the Solar system. We have just had a look at stray planets within our own system though, strictly speaking, rogue planets might be a better designation since Apollo objects do not stray in the accepted sense of the word. They merely pursue highly eccentric orbits which intersect with those of other planets including that of Earth. It is highly probable that in many other solar systems deep in interstellar space a roughly analogous state of affairs exists. However, when we speak of stray planets beyond the Solar System it is not these we mean but bodies of planetary dimensions, with or without satellites, but otherwise alone and *not* forming part of a solar-type system. The reader will at once see that this in effect takes us back to something of the conditions outlined in Well's "The Star" and Wylie and Balmer's *When Worlds Collide*

The first point we have to consider is just how such stray planets could come to be. Virtually all modern theories envisage stars and planets being created as a consequence of condensation from a common cloud of interstellar gas and dust. The pattern of any such condensation of interstellar material is, in the main, determined by the initial mass of that material. If this happens to be in excess of 0.02 of the Sun's mass, then at a certain point during the process of gravitational compression, the temperature in the central regions of the interstellar cloud rises to a point where energy-producing reactions occur, resulting in the formation of a star. If a small mass, one *less* than 0.02 of the Sun's mass, the thermal process is insufficient to attain this state. Under these circumstances a *planetary body* is created and *not* a star.

Some Russian astrophysicists (V. L. Strelnitsky and others) already claim to have discovered bodies which might conceivably represent an aggregate of protoplanets (planets not yet entirely formed) in the vicinity of a massive protostar. They believe that the observed dispersions of the radial velocities of these bodies indicate that some of the protoplanets are near to attaining the escape velocity of the system and could therefore be moving out into deep space. If these protoplanets were still insufficiently condensed, the material of which they are composed might simply disperse. If however the process of condensation has proceeded to a sufficient degree this could continue to completion giving either single or small groups of stray planets. It is also conceivable that the dispersion velocities of such a proto-system could be due to two factors—the pressures of radiation and the so-called "stellar wind" from the central protostar. The latter is probably the more important so far as the attainment of escape velocities is concerned.

But could the process of condensation be carried through to completion in protoplanets removed from the sphere of influence of the parent star? Astrophysicists tend to believe that complete condensation *could* be achieved, the more so if the process had already started prior to ejection from the system.

Since our galaxy has been in existence for a vast period of time, and assuming that the formation of stray planets can and does take place, then presumably many such objects must now exist within the great abysses which separate star from star. By making certain positive assumptions it has been suggested that the nearest stray planet to the Sun might be only three to four light years distant.

There is an alternative method whereby stray planets could be created. Stars which rotate slowly are believed to do so because they have lost much of their angular momentum to accompanying planets, i.e., in casting off material later to condense into planets. This, of course, envisages a slightly different form of planetary evolution. A few paragraphs back we spoke of interstellar gas and dust clouds condensing into stars and planets via protostars and protoplanets. In other words, each was forming contemporaneously and independently from the same mass of material. In the alternative process we are now considering, a protostar would first form and this would then give birth to planetary offspring by casting off chunks of its substance and, in so doing, a large measure of its angular momentum as well. Certainly there would seem a possible validity in this hypothesis, for when we examine the Solar System we find that the Sun, while possessing 98 percent of the system's mass, has only 2 percent of its angular momentum. The Sun is undoubtedly a slow-spinning star and it is a fact that all stars of a type and age reckoned to have possible planetary systems are slow spinners. Just what should be read into all this is still a matter for conjecture.

Noted British astronomer and cosmologist Fred Hoyle and others have shown that *further* angular momentum could have been shed by protostars casting material into deep space in order to attain their present observed states of dynamic equilibrium. Such material, if already partly condensed, could form stray planets. Eminent American cosmologist Harlow Shapley has stated that there could be a large number of such "orphaned" planets, probably ten per star.

Could stray planets be discovered either by observation from Earth or by orbital or lunar-based instruments? At the present time it is felt that this would not be possible from the surface of our planet. It is considered, however, that a survey site on the Moon's far side (the one permanently concealed from us Earth-bound mortals) might offer distinct advantages, especially if the infra-red portion of the spectrum were to be scrutinized. The infra-red flux (i.e., intensity of infra-red radiation) due to thermal radiation from a stray planet lying some three to four light years distant from Earth would not be detectable by present day instruments working from the surface of our planet. It could however be possible to achieve this from points on the Moon's far side. Should therefore at some future time an object be detected which has no visible companion star, shows high proper motion, proves to be a pulsating, coherent radio frequency or infra-red source and possesses large parallax, the possibility of its being a stray planet should not be overlooked. We would emphasize, however, that only when something of these conditions appear in the opening paragraphs of a science-fiction novel should we await (fortunately in imagination) the Apocalypse and Doomsday. On the other hand even a one in ten million chance still constitutes a possibility.

The chance of a properly constituted orthodox major planet, a member of a settled solar system, breaking free from the restraining hold of its parent star seems very remote indeed. Certainly the further out a planet lies the weaker becomes that hold. Pluto, five and a half light *hours* distant from the Sun is still held and controlled by it. Yet that hold is much less than that exercised by the Sun on Earth, which is only eight light *minutes* distant. Thus a planet controlled by but very remote from its central star would be able to break free much more easily. Nevertheless some external agency, force, or influence would presumably be necessary to provide the essential slight "push." This might just be achieved by the close passage of another star. If such a passage were reasonably close to a loosely bound outer planet then presumably that planet would only change masters. If, however, the passage caused a gravitational attraction only slightly in excess of that of the legitimate parent star, the effect might just be to cast the planet free as it lay briefly in a "null" gravitational zone due to the pull of one star cancelling out that of the other. But, as we have said before and must say again, the chances of two stars approaching sufficiently close to render such events possible are so remote that they can almost certainly be ignored. We say ignored rather than eliminated since, as we saw in chapter 3, the anomaly of Pluto might just have been due to such an occurrence.

And so now, rather briefly perhaps, to the remaining part of the theme of this chapter—that of artificial planets or planetoids. Since this aspect at once introduces the concept of highly advanced supra-intelligent alien beings, it may seem to belong more to the realms of science fiction than of science fact. This may not really be so but at the moment we have no categorical evidence to indicate that such life exists. The biological or life aspect as such is dealt with briefly in chapter 11. In the circumstances it seems unwise to make too much of the idea of artificial planets—guided, interstellar worlds. On the other hand the concept seems worthy of a brief mention.

The overwhelming difficulty about interstellar travel is the sheer distance involved. If extra-dimentional techniques to afford a practical short cut should prove impossible, or if they are only possible to a few advanced cosmic societies, the thing, if it is to be done at all, must be done the hard way. At velocities below that of light, time of travel must far exceed the lifetimes of living beings anywhere. For a long time one of the only few practical concepts has been that of the "space ark," the crew members of which would be resigned to living out their lives on board. It would be their descendants who would (hopefully) reach journey's end. Various designs have been suggested for such a craft but, in 1952, Dr. L. R. Shepherd of the British Interplanetary Society proposed that, instead of creating a vessel out of metal, it might be feasible to utilize a natural asteroid that had been hollowed out.

Admittedly to us this must be regarded as twenty-first or twenty-second-century stuff but to intelligent beings among the stars, members of civilizations much older and infinitely more advanced than our own, this could be both contemporary and very practical. Terrestrial men are already considering asteroid probes as a means to this end, though certainly not to use tomorrow or the day after. It is fairly certain that asteroids will abound in most, if not all, other solar systems as well so at least the supply of "raw material" should present no undue problems. Since many asteroids are presumably of nickel iron these would be mined. Thereafter a suitable small or moderate sized asteroid could be hollowed out. A civilization having the technology to mine asteroids should certainly be capable of this.

Propulsion might be achieved by means of an ion drive using a low boiling point metal such as cadmium or caesium. This combination would yield very high velocities over a protracted period with relatively low "fuel" consumption. Even so, the storage area for this "fuel" would have to be very large. Initial thrust would certainly be small but high velocities could be built up gradually. Such a travelling planetoid would obviously require a closed-cycle ecology with crops (and perhaps animals) being raised for food. It can be seen therefore that an asteroid "space ark" would have to be of considerable size.

An alternative of course, is a planetoid constructed more conventionally out of metal. This would probably be spherical in form and fashioned from material strong enough to render it immune to all conceivable onslaughts.

In essence this is about all we can or should say in these pages on the concept of "artificial" planets. They are a possibility, just how great we cannot say. The imponderables are too many and too great.

We should not then necessarily regard the tremendous gulfs that separate star from star as total voids. The possibility of natural, stray planets ejected from forming solar systems is a reasonable one. That of artificial asteroids is probably much less so but one that should still be borne in mind. The late H. G. Wells in his opening paragraphs of "The Star" wrote, "Beyond the orbit of Neptune there is space vacant so far as human observation has penetrated, without warmth or light or sound, blank emptiness for twenty million times a million miles. That is the smallest estimate of the distance to be traversed before the very nearest of the stars is attained." Perhaps when he included the words "so far as human observation has penetrated" he had decided even then, some eighty years ago, that it might be as well to leave some options open. We still should.

✠ir to Breathe

IN THE CHAPTER following this we will be taking a short look at the biological aspect in respect to extra-solar planets. This, of course, covers a very wide spectrum—virtually everything from the humblest microbe or plant spore to the most advanced technological race. Throughout this galaxy and the infinity of others it is very probable that all are represented, though our chances of coming into direct contact with any remain very slim, at least for the foreseeable future. There are, no doubt, those to whom such tidings come as a matter of profound relief, as well as many to whom they must be a source for regret. Both parties should perhaps bear in mind that certain of our more advanced cosmic brethren might just be able to seek *us* out.

So far in our review of other worlds we have paid scant attention to the matter of the atmospheres (if any) prevailing on these planets. This is something of great interest in itself and, of course, of profound importance in respect to life initiation and development. It is virtually impossible to conceive of any form of life, even the lowest, which does not require some sort of gaseous environment in order to exist. Admittedly here on Earth fish live in an aqueous environment though the vast majority breathe oxygen dissolved in the water by means of their gills. The fact remains that living beings require an atmosphere. Even our science-fiction friends would be hard put to get around this one—though no doubt there are those prepared to try.

In the first instance let us give some thought to the atmosphere of our own planet, not just what it comprises or how we make use of it, but how it got there in the first instance. We are not suggesting that our world is necessarily archetypal. So far it is just the one that we happen to know most about. It may thus serve as something of a prototype for many other planets of similar mass, constitution, and age that happen to orbit a G type star and lie at a similar distance from it. That distance incidentally represents the center of the ecosphere, a zone something less than ten million kilometers wide. Recent work by Michael Hart of NASA's Goddard Space Flight Center, as we shall presently be seeing, has indicated that the acquisition by our planet of a benign, life-sustaining atmosphere did not take place as a mere matter of course. On the contrary the route may have been a rather precarious one, since slight changes in any one of several factors in the past could so have changed conditions that the evolution of advanced life would have been rendered impossible.

It is by now generally accepted that the atmosphere of our world is, in fact, a secondary one. Shortly after its formation (relatively speaking) Earth lost its primordial atmosphere which was certainly very different from that obtaining today. Our current atmosphere may largely be due to material degassed from the interior of the Earth by volcanoes. This is so important an aspect that it should be examined at some length. Certainly the creation of the terrestrial atmosphere, as we know it, was certainly not a "one time" event in the remote geological past, but one which is continuing to this very day.

Terrestrial gases presently reside in the atmosphere, the hydrosphere (oceans, seas, rivers, lakes, etc.), and the lithosphere (the rocky crust). It is interesting to relate their presence there to volcanic activity, i.e., the upward movement of magma or molten rock from the lower crust or upper mantle of the Earth.

For a long time there was, perhaps understandably, a tendency to regard all volcanic action as being of a fairly uniform nature and the tremendous outpourings of vapor, so characteristic of major eruptions, as of little real consequence. It must be stressed, however, that volcanoes are variable in a number of respects, notably with regard to their distribution on the Earth's surface, the type of eruption and the nature of their solid ejecta (ashes, lava, etc.). Volcanic gases cannot therefore all be expected to contain the same elements and compounds. Neither should the amounts of these gases be regarded as inconsequential. Even when a volcano is *not* in eruption which, generally speaking, is most of the time, there is nearly always a certain amount of fuming activity. This should not be confused with *fumarolic* activity which is the large scale emission of gases, steam, etc., from vents and fissures in the opening and closing stages of many volcanic eruptions.

Our relative ignorance concerning the precise chemical composition of volcanic gases is due in no small measure to the undeniable fact that the procurement of samples is extremely difficult. Indeed at times it can only be described as thoroughly hazardous. Even when obtained, such samples are very likely to be contaminated by air or water vapor. And just for good measure there is every chance that the composition of the erupted gas will vary with the temperature of emission, and that there will also have occurred substantial losses due to sublimation (i.e., direct transition from gaseous to solid phase without passing through that of liquid) and to precipitation. Since volcanic sublimates tend to be rich in sulphur and halogen compounds (those of chlorine, fluorine, bromine, and iodine) it is perfectly valid to assume that volcanic vapor collected for analysis is deficient in these components. The final analysis is also difficult to adjust in the light of this deficiency because of the difficulty in estimating the amounts of these substances. As can be readily appreciated they are dispersed widely throughout the volcanic structure, and to some extent they are also dissolved and lost by the action of surface and of ground water. In these circumstances an estimate of volcanic gas constituents tends all too easily to become a "guestimate." However since volcanic gases at high temperatures are less likely to be contaminated by atmosphere and ground water, this fact offers a measure of hope. We find therefore, that samples are generally taken as gas escapes from molten lava at temperatures in the region of 1,000°C. The hazard aspect referred to earlier now becomes very plain indeed. Volcanoes are not noted either for their predictability or their docility. Approaching a stream of red hot lava issuing from an erupting volcano in order to secure a sample of emitted gas is not guaranteed to render the person concerned a good insurance risk!

This does not unfortunately constitute an end to the problem, for such a sample merely represents the gases emanating from *one* volcano at a particular point in time and from a single source in the vast world-wide system of volcanism. Since we are concerned with the outflow of volcanic gases into the terrestrial atmosphere over the past 4,500 million years or so we cannot claim that a sample taken in the manner just described is truly representative, either spatially or temporally. In consequence it cannot really tell us very much, if anything, concerning the development of this particular planet's atmosphere. Indeed it is true to say that attempts to follow the evolutionary pattern of the terrestrial atmosphere have been rendered very difficult in several ways, of which the above mentioned is but one. We can also mention ocean formation and reaction of atmospheric gases, interaction of gases with rocks, fluctuations in the Earth's albedo (reflectivity), "greenhouse" effect due to increasing amounts of carbon dioxide and the steady increase in solar luminosity with time.

The use of computers has, however, helped materially in this respect and scores of runs postulating many differing initial conditions have indicated that the present state of the terrestrial atmosphere (as well as that of its climate and surface) are capable of explanation as a consequence of the effect and interrelationship of the above mentioned processes. It is noteworthy that no extraordinary occurrences need be postulated. It would appear then that, generally speaking, our atmosphere is essentially orthodox in character. Therefore what has happened here could presumably happen also on other planets of a similar type, both in this galaxy and in others. Nevertheless we must at all times be careful not to attach undue significance to this. It represents a possibility and no more, an indication, perhaps even a lead, but certainly not a fundamental tenet.

According to the computer the best "fit" leading to the current terrestrial atmosphere began with a primordial atmosphere of 84.4 percent water (as vapor); 14.3 percent carbon dioxide, 1.1 percent methane and 0.2 percent nitrogen as well as traces of ammonia and the inert gas argon. It is of interest to compare these figures with those for the tholeiitic basalt volcano Kilauea on the island of Hawaii, viz. 64.3 percent water (as vapor), 23.8 percent carbon dioxide, 0.16 percent chlorine, 1.60 percent nitrogen, 9.97 percent sulphur, 0.05 percent hydrogen. Andesite lava volcanoes bordering the Pacific Ocean produce more water than this. We can at least see from these figures how known large scale and very extensive volcanic action could have been responsible for the high concentrations of water vapor and carbon dioxide in the primordial atmosphere. Such an immense supply of water vapor and carbon dioxide resulted after 150 million years in an unbroken cloud belt that shrouded the entire planet (comparison with present day Venus should be noted). Rising pressure caused much of the water to condense thus forming the first primitive oceans. Due to the large concentration of carbon dioxide a runaway "greenhouse" effect began and temperatures rose steadily. This was eventually countered as much of the carbon dioxide dissolved in the oceans to form carbonates (and much later hydrocarbons, as small creatures died, sank to the sea beds, and were covered over).

It was at this point that the first primitive life forms began to evolve in the early oceans. The waters contained large concentrations of dissolved carbon dioxide and ammonia where it was protected against the strong ultraviolet rays from the Sun. After about 800 million years the mean surface temperature probably reached its maximum (estimated at 317°K) and the pressure 1.4 atmospheres.

At this point in the proceedings the terrestrial atmosphere is thought to have been opaque to low frequency (i.e., long wave) radiation, thus preventing loss of virtually *all* the heat absorbed from the Sun. The "greenhouse" effect was now on in earnest. Evolving life forms set about releasing free oxygen into the atmosphere. This at once started to oxidize carbon. Thus organic carbon began to be deposited in conjunction with the normal sedimentary rocks of the period. After approximately one billion years an atmosphere had evolved which consisted almost exclusively of methane ("marshgas") and other chemically related hydrocarbons. At 1.7 billion years the continuous thick cloud enveloping Earth began to disperse. At once the "greenhouse" effect declined and temperatures, as might be anticipated, fell rapidly. Pressure also fell sharply to 0.6 atmospheres and the quantity of carbonate rocks (e.g., limestone, etc.) reached a level not markedly different from that of today. At approximately 2.45 billion years the remaining reducing gases were consumed, thereby rendering the terrestrial atmosphere oxidizing.

Geological evidence for all this is fairly convincing when we examine some of the oldest rocks. We can, for example, point to the presence of uranium oxide (uranite) and lead sulphate (galena) in sediments formed when Earth was some 2.0 to 2.5 billion years old. These clearly indicate deposition in a reducing atmosphere. On the other hand, thick sedimentary beds formed when Earth was 2.6 to 2.7 billion years old contain large quantities of red iron oxide, clearly indicative of an oxidizing and therefore oxygen-rich atmosphere.

It would seem that Earth's atmosphere at 2.5 billion years comprised nitrogen almost entirely, but thereafter a slow build up of free oxygen began. Mean surface temperature fell to 280°K and ice sheets appeared which eventually covered more than 10 percent of the terrestrial surface. Coincidentally cloud cover declined to less than 30 percent.

By about 400 million years ago oxygen and ozone (O_3) in the atmosphere constituted a fairly effective ultraviolet barrier. Freed from the inimical effect of excessive ultraviolet radiation, life, hitherto present only in the oceans, began to come ashore where plant life was already flourishing. Oxygen levels increased quickly.

It is postulated that two major crises took place during our planet's remote past, and after 0.8 and 2.8 billion years respectively. During the former, Earth only narrowly escaped an escalating, out-of-control "greenhouse" effect. Had Earth been situated only about five million miles closer to the Sun the temperature would have been such that water vapor would not have condensed, no oceans would have formed, and an extremely dense atmosphere essentially of carbon dioxide would have formed. This to all intents and purposes is presently the position with respect to Venus, which lies approximately thirty million miles closer to the Sun than does Earth.

At the time of the second crisis, ice caps had already reached out to cover more than 10 percent of the Earth's surface. Our planet was thus in danger of rapidly escalating glaciation— within a mere 1 percent of it in fact, according to computer simulation. Indeed had Earth been only about one and a quarter million miles farther from the Sun, these self same ice sheets would almost certainly have continued to spread, thus precluding the evolution of higher life forms. Today we look to Mars as an example of this kind of thing. Instead we find dry, dusty deserts rather than encroaching ice fields. But if, early on as seems likely, Mars lost most of its water vapor on account of its low gravity, then this is hardly very surprising. It has been seriously suggested that under the contemporary Martian surface vast ice sheets may lie. While not wishing to prejudge the issue, this does seem somewhat improbable. In the circumstances those who hold that a Martian race once existed are going to be disappointed. Mars, it would appear, could never have achieved anything approaching advanced life forms. But, if the first terrestrial explorers excavating part of the Martian surface discover even the most trivial indigenous artifact, then all that has been said and written in this vein will have to be very seriously revised. Science fiction writers will then be able to get busy on their tales of life, love and adventure in the long past halcyon of Martian civilization.

It is becoming increasingly clear however, that the habitable zone around a G type (i.e., solar) star is considerably less in extent than what had hitherto been supposed. The continuously habitable zone appears to be, relatively speaking, very narrow indeed— probably in the region of 6 to 6.5 million miles. This considerably reduces the estimated number of planets that have conditions appropriate to the evolution of advanced life in the galaxy.

Here then briefly is an account of the probable evolutionary course of events which gave Earth its present atmosphere. Could that atmosphere change again? We do not know. At present it is maintained by plant life since, due to the process known as photosynthesis, plants are able to absorb carbon dioxide during the hours of sunlight, while at the same time emitting oxygen. Without plant-life therefore, the future for oxygen-breathing, carbon dioxide-exhaling living beings on Earth would be, to say the least, highly uncertain.

We are perhaps reasonably entitled to expect some terrestrial type planets orbiting solar type stars to have atmospheres which followed a similar evolutionary pattern. When however, we think in terms of non-terrestrial type planets orbiting solar type stars, the position is almost certainly going to be different. This is merely stating the obvious, a fact made very plain when we consider the thick and (to us) poisonous atmospheres of Earth's sister planets, Jupiter, Saturn, Uranus, and Neptune. And this, of course, applies to, or could easily apply to, terrestrial type planets orbiting stars of a different spectral category. For example, had our Earth orbited at the same distance from it as an F type star, then the circumstances outlined earlier would have been realized. Water vapor due to the higher temperature would not have condensed, the oceans would not have formed, and the atmosphere would have remained thick, dense, and probably carbon dioxide-laden. But Mars would perhaps have profited and become much more Earth-like. In one respect we must be careful with this Earth/Mars analogy. Mars, being smaller than the Earth, has a lower gravity. The planet might have acquired an Earth-like atmosphere but the chances are that it would have been lost as a result of this factor. This probably happened and most of the oxygen now remaining on Mars is locked up in the rocks as red oxides. However an atmosphere evolves and, whatever constitution it possesses, it is still going to be dictated ultimately by a planet's mass and therefore the strength of that planet's gravitational grip on the atmosphere.

The overall picture that emerges, albeit dimly, is that only a fraction of the host of planets set around a myriad stars in our galaxy will have atmospheres akin to our own. Even in those the relative porportions of the two principal constituents, oxygen and nitrogen, could, and probably do, show distinct difference. On Earth the proportions are oxygen 21 percent and nitrogen 78 percent. If the oxygen content were somewhat greater and that of nitrogen lower, representatives of our kind could probably survive very comfortably, although too high a proportion of oxygen would probably lead to an induced sense of well-being and other complications. A reversal in these circumstances with nitrogen in higher proportions than on Earth could probably be endured for a time, though any pronounced insufficiency of oxygen could not be other than detrimental. Persons suddenly transported from normal altitudes to, for example, Quito, capital of Ecuador in the high Andes, are able to testify to effects of this nature, though after a time acclimatization seems to set in.

It is, of course, extremely unlikely that either ourselves or our immediate descendants will be called upon to endure atmospheric rigors on extra-solar planets. Interstellar travel will no doubt in the fullness of time arrive on the scene. That day however belongs to an era in the distant future. Long before then, we will somehow have come to terms with planets and moons in the Solar System that have either no atmosphere (e.g., the Moon) or nitrogen and carbon dioxide atmospheres (Mars, Venus). In "airless atmospheres" of course, if you will excuse such a contradiction in terms, there can only really be one answer—the pressurized, air-filled domes so beloved of science fiction writers. Since we are already able to fly in the near-stratosphere in pressurized airplanes this should not prove impossible.

When we come up against atmospheres such as that on Venus the problem is much less easy. This is something which may confront us a few decades from now in respect to Venus itself, or several centuries from now in respect to Venus-like atmospheres on planets of some of the nearby stars. This brings us slap up against the concept of advanced environmental engineering, or put in more homely terms, bringing about permanent and desirable changes in the atmosphere of a planet. It may be that intelligent races greatly in advance of our own, having achieved practical interstellar travel, are confronted with harsh Venus-type atmospheres on planets orbiting solar type stars. How may this seemingly miracle be brought about?

First of all let us take a close look at Venus, seeing it not only as a planet of the Sun but as a prototype for many planets within the galaxy. This seems reasonable enough. Already we have considered Earth as a prototype for planets of similar dimensions, orbiting stars of similar spectral type at roughly the same distance from their respective central luminaries. Just as there will be other Earths so there will be other Venuses. In chapter 2 we did consider Venus briefly describing it as the "hell planet." The closer look we now propose will surely endorse such a description.

Venus, it will be recalled from chapter 2, is approximately the same size as the Earth. In this respect they are sister worlds, but here all resemblance to one another ends. Some 90 percent of the planet's atmosphere is carbon dioxide. This may be directly attributed to the fact that the planet is so much nearer to the Sun than we are, twenty-six million miles nearer in fact. Carbon dioxide has, therefore, been able to accumulate in the atmosphere whereas on Earth it is largely locked in carbonate rocks. The origin of this carbon dioxide on Venus is almost certainly volcanic. As a consequence the greenhouse effect has simply "taken off." Ample confirmation of the resulting thermal effects has been supplied by both American and Russian spacecraft. The temperature on the surface of Venus is about 480°C (nearly five times as hot as boiling water). In fact it is possible that some areas on Venus may, quite literally, be red hot!

This is not the only effect attributable to the dense carbon dioxide atmosphere. It is also responsible for very high atmospheric pressures on the surface of Venus. Thus any astronaut landing on its surface who somehow escaped being cooked by the heat or asphyxiated by the carbon dioxide would be subject to an atmospheric pressure about a hundred times greater than on Earth. It seems unlikely he would wish to stay.

We are not quite finished with the nasty features of Venus yet. Very high atmospheric density would most likely lead to refraction of light rays to such an extent that our presumably far from happy astronaut would feel he was standing in the bottom of a deep bowl surrounded by high cliffs. In fact he would really be standing on a plain, the "high cliffs" being merely the horizon. How long would sanity last in such conditions?

Venus was, quite simply, created too close to the inner edge of the ecosphere and, due to excess of heat, has gained excess of atmosphere. It is unlikely that this was always so. Chances are that in the beginning the environments of Venus and of Earth were almost alike—both following a similar evolutionary path. But a billion years after the creation of the two worlds something began to go terribly wrong on Venus. Due to its closer proximity to the Sun, carbon dioxide started to accumulate in the atmosphere of the planet. This did not happen in a rush. At first it was a relatively slow process but as the millenia passed it began to accelerate. Today, some four billion years later, we see the frightful result. And although we speak here of Venus, the same effects as well as their dreadful consequences must have been repeated on many a planet deep within the stardust of the Milky Way.

Could visitors or intending colonists really do anything practical to alter such a state of affairs on another world? Until recently many might have regarded this as quite impossible. However Carl Sagan back in 1961 came up with a scheme staggering in the immensity of its scope, yet amazingly simple and straightforward. Sagan concluded, quite correctly, that the adverse conditions on Venus (and of course on all similar worlds anywhere in the universe) could be attributed to one single factor—the thick, carbon dioxide atmosphere. He suggests that the situation might be transformed if the carbon dioxide molecule could be split up into its two components, carbon and oxygen.

We know this takes place in nature as one of the stages in the process of photosynthesis. This process was mentioned earlier, and is the one whereby plant life treats the carbon dioxide we all exhale, resulting in continual replenishment of our oxygen. Indeed were there no oceans to supply the rain needed by plants we would soon have no oxygen to breathe. Sagan's scheme is therefore to introduce some kind of suitable plant life to Venus and allow it to break up the abundance of carbon dioxide there. The scheme is so delightful in its simplicity that we are immediately inclined to doubt its practicability.

Sagan is not, of course, suggesting that we deluge the Venusian surface with terrestrial flora carried to the planet and released by descending space probes. Such flora could neither root nor survive in the conditions prevailing there. What he does suggest using are blue-green algae. These organisms are neither wholly plant nor wholly animal. In dimensions they are microscopic. They also reproduce with great rapidity and are highly resistant to radiation.

Around three billion years ago the terrestrial atmosphere was one of carbon dioxide, ammonia, and methane. From the shallow, sunlit seas of that time came the blue-green algae which attacked the carbon dioxide in order to secure the carbon inherent in the glucose and carbohydrate food they required. The oxygen thus liberated eventually removed all ammonia and methane by a series of chemical reactions. Because of the presence of oxygen, animals were able to evolve. Since these exhaled carbon dioxide, more plants were then able to exist.

Sagan reckons his scheme would only take a few years to achieve the necessary results. Though such a time scale seems a shade optimistic, its simplicity and relative cheapness can hardly be denied. A few score of unmanned space-craft in criss-cross orbits around Venus would, every ninety seconds at 500 mile intervals, fire a rocket into the planet's thick, opaque atmosphere. In the nose-cone of each would be a colony of blue-green algae. On entering the atmosphere a small explosive charge would be fired releasing the algae which would at once start to feed and reproduce. Reproduction, once begun, is seen as very rapid indeed, which is why Sagan believes that in a year or two it would be possible to see at least part of the Venusian surface for the first time through the largest Earth-based telescopes.

Doubts concerning the ability of blue-green algae to survive and proliferate in an environment of almost pure carbon dioxide were dispelled in 1970 when several biologists, in a series of controlled experiments, proved that blue-green algae would survive. The tests involved the creation and use of a simulated Venusian atmosphere, pressure and all (or at least as much pressure as the walls of the containing vessels would stand without bursting). During these tests the algae began to produce oxygen at a prodigious rate, which, moreover, showed continuous increase. Indeed in one experiment each million cells opf algae increased the oxygen concentration by 380 percent *per day*! The most prolific of the blue-green algae used in these experiments was a variety found in hot springs. Certainly if this rate of oxygen production could be achieved and maintained throughout the entire atmosphere of Venus great, indeed spectacular, changes could not be long delayed.

Let us assume that Carl Sagan's plan has been put into effect. What would be the likely sequence of events? As the concentration of carbon dioxide falls, the infra-red radiation from the Sun hitherto trapped by the cloud layer begins to escape into space and the temperature of the lower atmosphere begins to drop very considerably. Water is then able to condense from the atmospheric vapor and fall as rain. But it would be rain the like of which we can rarely imagine. In comparison the Indian monsoon would seem like a summer shower. It has been estimated that sufficient water vapor to give one hundred inches of rain over the *entire* planet exists in the Venusian atmosphere. On to a glowing or baked surface that has never before known moisture would descend a cascading torrent of unimaginable proportions. On the first few occasions the rain might not reach the ground, being vaporized before it could do so. This, nevertheless, lowers the ground temperature. The next deluge, in like manner, lowers it still farther. Eventually the deluge reaches the ground. Deserts become lakes and the dust of untold aeons a mere turbid constituent of madly swirling rivers. Water soaks deeply into the ground (in the circumstances we can hardly use the term "earth"), thus preparing it for the coming of the complex molecules from which plant life will grow.

When rain finally reaches and drenches the surface of Venus the temperature drops to around 80°F, the equivalent of a hot summer day here on Earth and cool enough to sustain hardy examples of the flora and fauna of our own planet. Depressions on the Venusian surface now become oceans and the normal evaporation/condensation cycle of their water provides rain to nourish plants in precisely the same way as terrestrial flora is nourished. And, of course, dissipation of the former thick atmosphere ends forever the peculiar refractive effects. No longer does the horizon appear like a great high ridge seen from the bottom of a bowl.

With the sky now clearing of clouds, other than the conventional type due to evaporation of sea water, there can be seen a sight which in all the long history of Venus has never been seen before—the Sun! As on Earth, atmospheric oxygen would be acted upon by sunlight to form an upper layer of that essential allotrope of oxygen known as ozone (O_3). This is essential to life forms on a planet since ozone absorbs much of the dangerous ultra-violet rays emanating from the Sun.

Thus Venus would become habitable for men and women from Earth. A new world with rich untapped mineral resources would be at our disposal. There would be living space for millions from an overcrowded Earth. There would be arable land in plenty. Indeed some of the most pressing problems facing Earth and its peoples today might be solved for all the foreseeable future. On these grounds alone the leaders of our nations should start giving thought to the project. Politicians, unfortunately, rarely seem to see very far ahead of their own noses. It is up to the scientists to ensure that they are made to look, see, and understand. And that might prove more of an uphill fight than the Venus project itself!

Colonists who go to Venus after the successful completion of the project would not, of course, have everything their own way. Pioneers and colonists never do. We have seen this truth in the opening up of the American West. Courage, resolution, and adaptibility were called for. History shows these qualities were not lacking. Life on Venus would certainly not be without its inconveniences. Though we are digressing somewhat from the subject of atmospheres, this nevertheless seems a good point to enumerate them. The first of these is almost certainly the fact that the planet rotates very slowly. Indeed 118 Earth-days elapse between each sunrise. This makes for a rather longish day. In fact each day and each night would last for approximately 60 terrestrial days. These two month nights would be extremely cold and would bring conditions similar to those prevailing in our Arctic regions. The two month days, on the other hand, would be tropical. And just for good measure the rotation of the planet is retrograde so that the Sun would rise in the west and set in the east.

Let us emphasize again that in describing the "terra-forming" of Venus we have, also, been describing possibilities applicable to extra-solar planets of the Venus type, of which we can be fairly sure there is an abundance in the galaxy. Perhaps alien beings similar to ourselves have already achieved these things with some planets. Our distant descendants may well do the same thing, may indeed have to, should our Sun fade, swell out or "run wild."

So far as the Solar System is concerned Venus is really the only planet which could be "processed for human habitation." The reader will probably think of Mars in this respect (and automatically all Mars-type planets in the galaxy). Now Mars today is very loosely speaking, much more Earth-like than is Venus. It might therefore appear an easier prospect for treatment. Mars, as we saw in chapter 2 is a much smaller planet than the "twins," Earth and Venus. As a consequence its hold on an atmosphere is considerably less. Thus, if we could give it an atmosphere in some miraculous way, we would need to keep replenishing this to make up for the continuing escape of gas molecules into space. Carbon dioxide in the atmosphere of Mars is roughly one ten thousandth that of Venus. There would be no point in seeding this with algae. In fact it would be advantageous somehow to make Mars warmer, for it is a very cold planet indeed. Clearly the oxygen available from this amount of carbon dioxide would be totally insufficient. Oxygen on Mars would appear to be "locked" in the rocks as red oxides. Unfortunately we know of no convenient organism which could digest these and release it. Recent evidence indicates that Mars did at one time possess a reasonable abundance of water and probably also a reasonable atmosphere. These things have long since gone, and unless terrestrial technology can make a great break-through they are going to stay gone.

It should be placed on record that Carl Sagan has also come up with a scheme for Mars but this entirely lacks the consummate beauty, ease, and efficiency of his plan for Venus. Sagan's plan in this instance involves melting a polar ice cap and distributing the resulting water and water vapor across the surface of the planet. As this plan unfolds we realize we have been here before for did not Lowell's Martians do this very thing, hence the legendary "canal system?" The scheme has great weaknesses. For a start, how much water is available from the pole caps? In this respect they cannot be compared with the vast ice fields surrounding the terrestrial poles. Low gravity would soon mean the loss of any water vapor into space. Low pressure of the Martian atmosphere (and it *is* very low) would soon take care of the liquid water also. Neither should we forget that such a scheme would involve vast engineering works. These would be prohibitively expensive and very difficult to set up and operate.

In the case of Mars (and all Mars-like planets) it looks as if we must revert to the pressurized domes of science-fiction. This is highly ironical when we consider how placid are the surface and atmosphere of Mars compared to those of Venus. So in the Solar System man is more likely to migrate to Venus than to Mars. Beyond, the Solar System he will prefer Venus-type planets than those resembling Mars. Naturally the type he will most seek will be the terrestrial variety. There will, in all probability, be plenty of those too. He may, however, have greater distances to cover before coming upon them. On our world minor changes in any one of several factors in the past could have changed conditions sufficiently to render evolution of advanced life impossible. For this reason atmospheres immediately appropriate to our kind could be at a premium in the galaxy. In the immense host of stars they *will* be there. This does not mean they will necessarily be close to us. This inevitably has a direct bearing on terrestrial type life, perhaps on life in general. It could be near. It could also be very remote. The latter is perhaps the more likely. This does not alter the fact that it exists. In the recent words of Carl Sagan, "Personally, I think it far more difficult to understand a universe in which we are the only technological civilization or one of but a few, than to imagine a cosmos brimming over with intelligent life." With these words echoing in our ears let us now in the next chapter think briefly about life on other worlds, its possibilities, and its probable forms. Were our Sun to die this is a factor which would have to be faced and faced squarely!

And Lo
There Was Life

THE PRIME PURPOSE of this book is to present a popular exposition on the subject of extra-solar planets, the possibilities of their existence, the form they may take and the whys and wherefores of their origins. The question of life on such worlds is one that generates a tremendous and an increasing amount of interest. Clearly it is a popular and highly intriguing one. In these pages however we prefer to keep this aspect to a minimum since the subject has already been extensively dealt with by a number of writers, including the present one. On the other hand the question must arise as to whether, in all fairness, we can justifiably ignore this aspect entirely in a work which purports to deal with all, or at least as many aspects of extra-solar planets as possible.

In the circumstances the wisest course seems to go for a compromise, to have a single chapter dealing exclusively with the biological connotations, but on this occasion to draw upon the latest beliefs and opinions on the subject. The reader may point out that references were made to conditions for life in those chapters dealing with the planets of multiple and of variable stars. This was done, however, to illustrate as fully as possible the rather extreme conditions which could prevail on some of the worlds of such stars.

The belief in the existence of life and of advanced technological civilizations on remote planets of other stars is one that has waxed and waned repeatedly over the years. During that period when the Solar System was seen as the consequence of a virtually unique cosmic accident the question barely arose. No other planetary systems anywhere—then no other life anywhere! It was as simple as that. At that time, of course, it was still just possible to believe in the existence of Martians, perhaps even Venusians, so we poor miserable terrestrials didn't feel too alone and deserted. But immediately the realization arose that a high percentage of stars could, and probably did, have planetary systems the pendulum swung violently in the opposite direction, to the infinite relief, we must imagine, of "sci-fi" writers to whom the demise of Martians and their ilk could hardly have been other than a severe body-blow. The parameters of time and distance were still an obstacle of course, but time and space warps could always be called in to obviate these difficulties. Indeed time and distance might even be regarded as friends for they precluded most effectively and throughout all the foreseeable future that scientists could once again desecrate the holy preserves of the fiction writers.

In the strictly scientific field, as we have already said, the concept has tended to wax and wane. There are those who would have life and civilizations spread throughout the galaxy like currants in a cake! Others again, for one reason or another, adhere to the belief that life, especially advanced intelligent life must be strictly limited. The truth probably lies somewhere between these two conflicting viewpoints, but just where remains a matter of some doubt.

During the last decade the tendency has probably been more in favor of the "currant cake" idea but at present the trend again seems to be swinging in the opposite direction. This may well represent commendable caution.

Until quite recently, estimates and predictions suggested up to a million advanced civilizations in our galaxy alone, and probably a like number in every other galaxy. And these, could one add them up, would amount to an awful lot of civilizations. It has always been accepted that life could only be initiated and develop in continuously habitable zones about a star. We see this quite clearly in our system of planets about the Sun. Only Earth qualifies and only Earth, to the best of our knowledge, possess life. A recent study of continuously habitable zones about various types of stars suggests that perhaps far fewer planets than previously thought have conditions appropriate to the establishment and evolvement of life.

Dr. Michael Hart of the Systems and Applied Science Corporation has extended to other stars his earlier study of the evolution of terrestrial conditions throughout the lifetime of the Sun. His conclusions, whether we choose to accept them or not, are interesting. He maintains that life only *just* managed to exist on Earth, pointing out that had the Earth's orbit been only 5 percent closer to the Sun (about five million miles), an escalating "greenhouse" effect some 3.7 billion years ago would have produced a Venus-type planet. And, alternatively, had Earth's orbit been a mere 1 percent farther out (about one million miles), runaway glaciation 1.7 billion years ago would have given us (only there would have been no "us"!) a Mars-like planet. Our sovereignty of the Solar System seems to have been a very close run thing.

Hart stresses that the same principles must apply to planets surrounding many stars. Estimates of the number of planets where life could have evolved have hitherto been based largely on the assumption that life would probably develop and evolve on any appropriate planet which had liquid water on its surface at some time. Hart points out that escalating "greenhouse" and glaciation conditions must inevitably reduce the zone of continuous habitability or ecosphere, i.e., that concentric shell around a star where conditions are sufficiently benign during the three to four billion years required for the emergence and development of advanced life forms.

Smaller, less luminous stars than the Sun must obviously have a narrower, less extensive ecosphere. This number, Hart maintains, would rapidly shrink to zero in respect to a typical class K1 star having a mass equivalent to 0.83 that of the Sun—thus *no* habitable zone around most K or M class stars!

Clearly the "life zone" must be wider and more distant from the central luminary in the case of stars having masses greater than that of the Sun. This might seem to "balance the books." Unfortunately, those having only a mere 10 percent greater mass would as likely as not emit sufficient, ultraviolet light after four billion years to hamper, if not to inhibit, the spread of life from the seas onto dry land. The fact must also be faced that stars greater in mass than 1.2 solar masses, though they avoid escalating glaciation after 3.5 billion years, are likely to remain too hot for some four hundred million years.

Hart's conclusion is that only stars having masses between 0.8 and 1.2 times that of the Sun are really appropriate to advanced life forms, and that not all of these could be expected to have a terrestrial-type planet conveniently orbiting in the relatively narrow zone of continuous habitability. All this does not, of course, mean that we are likely to be alone in the universe. What it does imply is that we may have rather fewer neighbors than is popularly believed.

In essence, Hart's conclusions do not really change the situation regarding the belief in extra-terrestrial life of an advanced kind. The tenets of his claims have been widely accepted for years and it would be difficult to dispute them. Indeed by emphasizing them so strongly, Hart may have done a very considerable service to the intriguing theme of extra-terrestrial life. He has really emphasized the need for keeping a tight hold on the reins. This is no bad thing, for the concept of extra-terrestrial life is one which can run away all too easily, especially when an excess of enthusiasm is the spur.

It cannot be truly said that Hart has given the question a pessimistic slant, as has been suggested or implied in some quarters. What he *has* done is to counsel caution and to help prevent excessive zeal taking the place of reason. It *is,* of course, possible, as all devotees of science fiction know, to speculate on life forms able to withstand far greater extremes of heat, cold or ultraviolet radiation than those on this planet. This, at best, always represents a risky line of argument. It is certainly possible that some forms of life could possess greater resistance to these things than do any here on Earth, but we would be most unwise to extrapolate this line of thought too far. Science fiction has, of course, given us beings based on alternative chemistries, e.g., silicon, and for this, for all we know, there could be a certain validity. It is essential however, that we adhere to a reasonably strict scientific discipline. The truth is that we have *no* experience whatsoever of such alternative biochemistries. The carbon-oxygen system on which living organisms on this planet are exclusively based still seems the most rational and the one most in accord with the fundamental laws of physics, chemistry and biology—laws which we must assume are universal in the fullest sense of the word.

When we consider the vast number of stars within our galaxy, it is clear that the sensible restrictions stressed by Hart still leave the possibility of advanced life within our galaxy at many points as a most distinct possibility. It may mean, however, that we or "they" might have to travel further to meet each other but this is only what Carl Sagan has been saying for years. His belief is that we might have to look out as far as 300 light years ere we came to another civilization. This admittedly inhibits the concept of interstellar commerce and relations unless, of course, sophisticated techniques of extra-dimentional travel can be devised and developed.

The balanced verdict today therefore would seem to be that we are *not* alone in the galaxy, that this galaxy and all the countless others probably abound with life forms of one sort or another but—and this is important—life will *not* be found in the environs of every other star.

Next we come to the second fundamental: what of the physical appearance of intelligent aliens on the planets of other stars? Here again, much has been spoken and written on this intriguing theme. It is, unfortunately, one on which we can, at best, only speculate. There exist, of course, two types of speculation, the sensible and the other. In the light of all that has gone before, it is probably best to continue the practice we have so far adopted, to update and that briefly.

Despite the many weird, wonderful and ofttimes downright repulsive aliens dreamed up by movie makers and "sci-fi" authors, the fact remains that the laws of biology, zoology, anthropology, exobiology—and common sense—come down in favor of a basically humanoid form with respect to intelligent life across the galaxy. And though we may be quite wrong, beings based on carbon make much more sense than any based on silicon. Silicon polymers of the protein type seem most unlikely to form the compounds essential to biochemical evolution. The energy requirements for a living system seem capable of fulfillment only by carbon in conjunction with the high-energy phosphate bond. The rock-eating silicon creature which flickers from time to time on our television screens probably belongs only there! And since it is extremely difficult to envisage life based other than on carbon compounds forming in water, this seriously restricts planets having intelligent life to those with broadly terrestrial-like surface temperatures and pressures. For another reason the *type* of star is also restricted, for the DNA (desoxyribonucleic acid) molecule is sensitive to high radiation levels, especially those in the ultra-violet region. In short any old type of planet or star will *not* do. So out too go the queer balloon-like reaction propelled creatures envisaged in the atmospheres of Jupiter and all other Jovian-type planets.

Note however that we have been referring so far only to intelligent creatures. Here on Earth we have some rather strange and certainly unique beings. Take for example the giraffe. Now if the giraffe did not exist on Earth and some one had dreamed it up for a "sci-fi" movie most of us would be extremely skeptical—and not perhaps without reason. Conversely it might be possible for nonintelligent life on alien planets to have odd forms simply because of a niche in nature they are designed to fill or, as in the case of our giraffes, because of some essential food-grazing characteristic.

Were this a volume dealing with possible life forms on other worlds only we could go into the matter in considerable detail. We could deal with the general anthropomorphic view, the idea of conceptualization and intelligence, predatory supremacy, locomotion and sensory organs. Since this is not our exclusive brief, we will simply summarize the latest views on likely extraterrestrial form.

1. *Structure:* Build and height of alien beings has often in the past been directly linked with a planet's gravity. For a high gravity planet we might (justifiably it seems) expect build to be squat and muscles to be extremely powerful. In the cases of a low-gravity planet the opposite might be expected to apply: tall, more fragile structure, less powerful muscles. The only snag about this argument, and for long it seems to have gone unnoticed, is that here on Earth, a very moderate gravity planet, we

have examples of both types. The truth of the matter is probably that there is a *range* of build and height parameters applicable to creatures and beings on specific planets.

Differences between the sexes of intelligent beings might not be as distinct as they are here on Earth. Much could depend on the form of procreation favored by nature. In 1973 a placque was specially designed for the Pioneer spaceprobe which will (if it has not already done so) leave the Solar System forever. This was designed by Linda Sagan, wife of Carl Sagan. One of the features it is designed to demonstrate to alien eyes, should it ever reach the inhabited worlds of another sun, is the distinct difference between the sexes of "Homo sapiens." This might not be under-stood if the alien beings concerned were less sexually distinct. Nevertheless in the case of mammalian creatures the consensus seems to be that terrestrial physiology is the type most likely to be favored.

2. *Color:* This is a rather difficult facet to consider. On Earth, color in respect to men and women ranges from white (which is really pink) through various shades of yellow and brown to black. Roughly speaking, pigmentation of the skin on our planet seems to bear a close relationship to geography, i.e., the hotter the climate the darker the skin. Heredity of course, has long since taken over so that by now genes mean more than geography. Thus a child born of black parents in northern Europe will be black, just as a child born of white parents in Africa will be white. Extra-terrestrial beings might show all the skin color variations we have here on Earth plus perhaps a few more for good measure.

3. *Facial Characteristics:* There are certain definite constraints here due to the organs of smell and taste being in close proximity to the mouth, the necessity for stereo vision (two eyes) and binaural hearing (two ears). Despite the occasional appearance in science-fiction of intelligent beings with one central eye, this seems a most unsuitable arrangement in respect to discerning perspective and depth, unless of course that single eye is of a highly specialized "bi-focal" type. Nevertheless two separate eyes seems a much more sensible arrangement. We can thus reasona-ble anticipate the general facial characteristics of alien men and women being on a par with our own, though obviously within the limits we have described there is scope for considerable variation.

The possibility, then, of what might appear to us a thoroughly grotesque caricature of human features cannot be discounted. At the same time of course, we must remember that the reactions of intelligent, cultured aliens to our features could be very similar!

4. *Appendages:* By appendages we mean hands and feet, or more correctly fingers and toes. Variations are, we must suppose, possible though we would imagine that an excess of fingers and toes would prove as great a handicap as too few. So whereas five fingers on each hand and five toes on each foot might be replaced by four or by six, this would seem to represent the limit of sensible variation if the full manipulatory potential of these organs is to be retained. At times in science fiction works we are treated to intelligent beings with, for example, an additional pair of arms. This seems improbable. Apart from giving such a being more than a passing resemblance to an octopus there seems little point in having extra limbs.

5. *Internal Organs:* Here we come into very unknown territory indeed. In considering the internal digestive, cardio-vascular and pulmonary system of the intelligent extra-terrestrial being there can be no certainty whatsoever that these would be in any way akin to our own. Possible variations are legion, certainly too many and too involved for discussion in a book of this kind.

This then is really all that can safely be said in respect of both the existence and the possible forms of intelligent alien beings on planets of other stars—or, for that matter, in other galaxies. Inevitably however, an element of speculation must creep in, but we would hope that the contents of this chapter represent rational and sensible guidelines.

It is difficult to believe that life in the universe is confined to Earth, a small planet of an insignificant star in a probably not particularly significant galaxy. By any standards the mathematical odds would seem overwhelmingly against such a possibility. It is far easier, and seems infinitely more rational, to accept the view that life in this galaxy of ours (and that includes intelligent life) has proliferated extensively. This does *not* mean that it is everywhere. Such a supposition would be equally irrational. The truth lies somewhere in between. Just where we cannot tell, but someday perhaps, we may.

⑫

Per Ardua
Ad Astra

T O REACH THE stars we must have vehicles capable of taking
us there. At the moment it is very difficult to envisage the
forms these great creations of technology will assume, the starships which will transport our distant descendants, freight and
other essentials to the worlds of other stars. In an era in which
mankind has only just reached the Moon this can hardly be other
than obvious. Yet like all other creations of the human race,
starships must have their genesis. Great sea liners like the *Queen
Elizabeth II* are, after all, directly descended from the coracles
and other frail craft used at the dawn of civilization. We see this
in aviation too, in Montgolfier's hot air balloon and the Wright
brothers' frail biplane to the giant super-jets of today. A lot
happened in between, great successes and disastrous failures. It
could hardly have been otherwise. In space, circumstances in this
respect will be no different. This we must be prepared to accept.

In the context of star ships, even starting at the beginning is
difficult. Right now none exist and are hardly likely to in the
foreseeable future, either. The Saturn 5 rocket which put men on
the Moon is probably the progenitor of the starship, though in
itself it is no starship. Where then do we find our genesis?

Probably the most appropriate place to start would be to examine in broad detail the findings of a serious study project carried

out on the subject. In February of 1973 the British Interplanetary Society in London decided to launch just such a project. A highly qualified team was assembled and given the rather daunting job of visualizing the Earth's first starship. It was mid-1978 before the mammoth task was completed. Over 10,000 man hours of work are estimated to have been spent directly on the project which was code-named Daedalus after the mythological Greek architect supposedly responsible for designing the famous maze on the island of Crete.

The mission profile studied in detail was for an unmanned fly-by of Barnard's Star which lies some six light years from the Sun and, after Alpha/Proxima Centauri, is the closest star to us. The vehicle, as it emerged in its final form (though not, of course, real existence) consisted of a two-stage nuclear pulse rocket capable of covering the intervening distance in approximately half a century. This may seem a tremendous transit time, but recall that the Wright brothers' biplane could hardly have been expected to keep pace with Concorde. In space too we must creep before we walk and walk before we run—and run for long before we fly!

The concept of a two-stage vehicle arose primarily from the shortcomings revealed by earlier studies involving a single stage vehicle. This was attributable largely to the problem of erosion by the interstellar medium of the single large engine of the spacecraft. Consequently the addition of a smaller engine and redistribution of the propellant represented a logical step.

The design which has emerged envisages a craft with a mass at engine ignition of approximately 54,000 tons of which 50,000 tons represents propellant. This constitutes a distinctly high ratio of propellant to payload, and constitutes what will surely be one of the first great problems in starship design and construction. It is reckoned that the entire mission would involve a twenty year period for design, manufacture, and final adjustment and check out, fifty years of transit time to reach Barnard's Star and some six to nine years for transmission of information back to the Solar System. Thus if practical application of the project were to be put in hand by say 1985 (highly improbable!), it would be around 2065 before any concrete results could possibly be achieved. Admittedly this may seem a most inauspicious beginning but we must never forget, never be unaware of the tremendous magnitude of star travel and all the awesome parameters of time and distance which are inevitably involved. We must be severely practical and utterly realistic—always.

The idea is for the vehicle to leave the Solar System in the vicinity of Jupiter. We are often tempted to think that any star ship will leave the Solar System by traversing its plane, finally crossing the orbit of Pluto, that planet serving as a kind of last frontier. This is all very poignant and romantic but must nevertheless be regarded largely as a piece of poetic license. In certain circumstances it could, of course, be true. The fact is however, that a starship bound for a destination in deep space leaves the Solar System just as soon as it sets course a few degrees above or below the *plane* of the system. It would certainly still be affected by the gravitational attraction of the Sun, and to a lesser extent by the lesser bodies of the Solar System. Nevertheless the ship would be *leaving* the Solar System. But only when finally and irrevocably clear of the Sun's attraction could it really be said to have *left*.

The boost period (involving three propellant tank drops and a stage separation) would be maintained for 3.8 years. Following this a coasting velocity of 12 to 13 percent of light velocity would be attained (about 23 thousand miles per second). During this coasting period the instrumented payload would be operative, measuring several of the parameters associated with the interstellar environment. We should remember that this project represents a pioneering mission in the fullest sense of the word. Whatever we may presently think to the contrary, interstellar space is still an uncharted void so far as travel through it is concerned.

On arrival in the immediate environs of Barnard's Star a dispersible instrumented payload would be released. In this respect something along the lines of a MIRVed intercontinental ballistic missile warhead is envisaged. Several components would separate to pursue their independent ways and purposes. For these particular operations the main propulsion system would be employed. Information gathered as a result would then be transmitted back to the Solar System by microwave transmitters.

The timetable for all these events and their sequence is as shown in Table 3.

Table 3

Event	Time After Launch (Years)	Distance from Sun (Light Years)
1 1st stage separation	2	0.06
2. Boost termination	3	0.20
3. 1st mid-course correction and commencement of coasting	4	0.25
4. 2nd mid course correction	25	3.00
5. Probe launch	46	5.75
6. Arrival	49	6.00

The payload seen as necessary for all these operations is estimated at 500 tons, a high proportion of this being represented by the dispersible payloads.

It might well be asked why Barnard's Star was chosen as the objective of the Daedalus craft since this star lies about twice as far distant from us as Alpha Centauri. It must however, be realized that the design of any interstellar mission relies on accurate data on potential targets so as properly to assess the objectives and direction of the mission. Much of the presently publicized astronomical data relating to the nearer stars tends to contain unacceptably large discrepancies. In the light of this fact it was considered highly desirable that a standard set of data be used in the project. This was duly produced and included accurate data on some ninety stars within twenty light years of the Sun. The data concerned involved distance, position, luminosity, magnitude, mass, radius, spectral class as well as details regarding the proper motion of the stars involved.

During the summer of 1963, as explained in chapter 6, Peter Van de Kamp, after many years of painstaking and exhaustive study, reached the conclusion that Barnard's Star was attended by a planet (or planets). So far it has been impossible to reach any such conclusion in respect of Alpha Centauri. Clearly it is of more interest to send an interstellar probe to a star *with* planets than to one probably devoid of them. The choice of Barnard's Star remains arbitrary however, and could be changed, since the capability of the vehicle for more extended missions (as well as lesser ones) was borne in mind. This is not to suggest that such capability is excessive. Indeed the "built-in" capabilities of Daedalus restrict it to a range of about ten light years and a mission period of a hundred years.

There was, however, more to the choice of Barnard's Star as objective than the mere planetary aspect. Life potential was also considered viz. relation to stellar evolution, possibility of autotrophic life forms (those obtaining nutrients from external sources, i.e., most animals). These factors were applied to the thirteen nearest stars, which included Epsilon Eridani, 61 Cygni, Epsilon Indi and Tau Ceti, because of their similarity in so many respects to the Sun.

Each star on the list was allocated a certain arbitrary classification or rank, the *lower* the rank, the more appropriate each star. The results worked out as follows (Table 4).

Table 4

Star	Rank	Star	Rank
Nil	1	Luytens 726-8	7
61 Cygni	2	Wolf 359	8
Barnard's Star	3	Epsilon Indi	9
Epsilon Eridani	4	Sirius	10
Tau Ceti	5	Lalande 21185	11
Alpha Centauri	6	Ross 154	12
		Ross 248	13

Missions to the five highest ranked stars were regarded as capable of providing a representative distribution of directions in neo-solar space, a factor from which interstellar properties and stellar parallax measurements taken during flight could most benefit.

Nevertheless one of the main reasons for the choice of Barnard's Star (Rank 3) was the likely existence around it of one or more planetary companions. It is perhaps unfortunate that during the very recent past some doubts have been cast concerning the findings of Van de Kamp in relation to Barnard's Star. Whether or not these are in any way valid is something which has yet to be decided. Van de Kamp himself admits that further observations and more time are necessary to "separate" the respective orbits of the possible planets of this star. It may be that his announcement in the early summer of 1963 was a little premature. At the time it did not seem so. He may still be right. But now back to the Daedalus star probe itself.

Prior to the commencement of the project study it had been accepted that the only valid high-performance propulsion system bringing interstellar potential within sight was the nuclear pulse rocket with external ignition. The Daedalus project envisages fusion reactions between the nuclei of light elements. This was regarded as preferable to the fusion of heavy element nuclei, which provide less effective energy.

Engine mass amounts to 500 tons, with a specific power of 100 megawatts per kilogram. Effective exhaust velocity is reckoned in the region of 10 million meters per second. Not surprisingly, engine design proved a considerable problem as did the specific nature of the propellants. Eventually deuterium (an isotope of hydrogen) and helium 3 fusion were decided upon. The reaction products in this case are all charged particles which would contribute to the thrust of the engine.

Small amounts of deuterium and helium 3 in the shape of small spheres a few centimeters in diameter would be injected into the center of a cusp-shaped magnetic field. On reaching the target they would be impacted simultaneously by high power electron beams. The spheres would as a consequence be worn away, due to the high thermal rates and energy deposition in these outer regions. This would lead to the generation of very high surface pressures, fuel compression and shock heating, the central regions attaining temperatures at which thermonuclear reactions would begin to occur. The resulting expanding plasma ball, being highly conductive, would sweep aside the magnetic field within the reaction chamber. Enclosing this reaction chamber would be a thin large diameter shell of molybdenum. Because of the rapidity of magnetic field deformation it would be most effectively confined and compressed between the conducting shell and the plasma sphere. The kinetic energy of the plasma would be temporarily stored in the magnetic field, reversing the direction of motion of the plasma, ejecting it at high velocity along the axis of the engine. Exhaust momentum would be transmitted to the shell of molybdenum and then, by means of thrust aperture, to the actual vehicle.

The propellant itself is seen as being carried in the form of preformed spheres within several drop tanks. These spheres would essentially represent a deuterium honeycomb full of helium 3, storage temperature being about $3°K$.

The foregoing paragraphs represent merely the essentials. Clearly there is a lot more to it than this but these few brief facts are probably all that is desirable or necessary in a book of this nature. For these facts I am indebted to the British Interplanetary Society of London, whose team took on what can only be described as a task of monumental proportions.

The form of the first true interstellar probe leaving Earth for the environs of another star depends largely on the shape of future development and discovery. One feels however that Earth's first endeavor in this respect may not differ all that greatly from the probe envisaged in the Daedalus project. In the fullness of time no doubt, such creations will find their way into museums, much as early aircraft, once the wonder and pride of their day have done.

Clearly the generation gap separating Daedalus and like probes from the great interstellar freighters and liners which will one day bear the seed of the human race from Earth and the Solar System to the worlds of other stars is enormous. To extrapolate in this direction is extremely difficult. If we could foresee the pathway and pattern of future technology the position would be easier, but that in itself is almost equally difficult. Will the starships of the future, because of the awesome parameters of time and distance, be merely one-way ships designed as "generation travel" or cryogenic ("suspended animation") vessels, or will some of the concepts outlined in my recent book *Interstellar Travel* come into vogue? Should the latter be so, then perhaps we can foresee the formation of interstellar passenger and freight services along the lines of contemporary air transport.

So in writing this chapter no effort will be made to foretell the intermediate stages, but we will try, no doubt imperfectly, to gaze into the far future, trying thereby to gain a hazy vision of the shape of things to come.

The concept of regular interstellar services is one that has intrigued and fascinated all science-fiction buffs for many decades now. This is thoroughly understandable so we will stress this aspect of interstellar travel rather than the sadder and more poignant concepts of the one-way ship bearing successive generations or occupants frozen in a state of suspended animation. But we must also remember that if, for any reason, our continued existence within the Solar System became impossible, for example by some unexpected turbulence on the Sun, our technology might have advanced no further than the one-way ship. But under such dire circumstances only one-way ships would be necessary.

The types of starship appearing in certain movies and TV spectaculars might, in some cases, not be too wide of the mark. We are thinking especially of those in the films *Star Wars* and *Close Encounters of the Third Kind.* In the second the alien starship was clearly of the "flying saucer" type but none the less impressive for that.

If a "trans-galactic" starship (to use the adjective beloved of science fiction) is to be what it purports to be, then it must be equipped with the means of "telescoping" space in one way or another. For this let us use the generic term, "warp" method. This could include travel via hyper-dimensional space, via a rotating black hole or the like, or some other amaxing system that I haven't quite figured out yet!

The essential feature which such a starship must possess is a very efficient and reliable "warp" guidance system. When a ship goes into "warp drive" (another good "sci-fi" expression) it is taking a short cut—one might say the short cut to end all short cuts if it manages to bipass a whole galaxy! But a short cut is no short cut if it happens to deposit you at a point several light years from the one aimed at. Thus the crew of an interstellar ship, whether a super liner of humble freighter, going into a "warp jump" will have to be very sure that their ship will wind up where it was intended to. If in a jet-liner today, bound say from London to New York, the captain were to announce that he had somehow got himself over Miami, the passengers would, not surprisingly, consider the man a fool and make a mental note never to fly that airline again. At the time they might be pacified by extra food and free drinks from equally exasperated stewardesses. But en route from Earth to Alpha Centauri it would be most highly disconcerting—and that is putting it mildly—to come out of warp in the vicinity of Sirius. The poor starship stewardesses might be kept rather busy on the free drink racket before this particular mess was sorted out. They would probably be tempted to have a few themselves!

In early days of "warp" travel (assuming this ever becomes a really practicable technique) a starship emerging from a warp jump might expect a *minor* degree of error. So long as it was only minor, and very minor at that, this would merely be a nuisance, involving perhaps a small amount (very small) of space to be traversed through normal space under conventional power, in other words a course correction. We would be inclined to assume however, that by its nature warp-jumping would entail a tremendous drainage from the power banks of the starship. This would probably have to be made good before completion of the remainder of the voyage through normal space under conventional propulsion. What is required is, therefore, a warp-drive system sufficiently sophisticated to permit the ship to emerge from hyper-space (the extra dimension) *very close* indeed to its destination star (not too close either of course—stars tend to be rather hot objects.) The distance requiring the use of normal propulsion would therefore be acceptably short. In the event of regular scheduled interstellar services such conditions would be mandatory.

Whether a ship utilizes warp jump through the frightening utter darkness of hyper-space or is a more conventional ordinary space "plodder," it must have a propulsion system. It may be appropriate therefore to look very briefly and as non-technically as possible at some of the possible options which may be open to us in the fairly foreseeable future, say next century sometime. What the options could be in the far future is just about anyone's guess. For the present the following are regarded as having definite relevance:

1. The nuclear-electric ion rocket:

This is essentially an ion rocket deriving its electrical power from a fission reactor. Such an arrangement might be expected to yield exhaust velocities in the region of 1,000 km/sec. The potential here is not very high since to attain a velocity of 0.1C (where C represents the velocity of light) a mass ratio of ten billion (10^{12}) would be required. This is a pretty hefty figure to achieve a velocity of 18,600 miles per second. For a velocity of 0.01C (1,860 miles per second) mass ratio drops dramatically to twenty or thereabouts. Unfortunately this would involve very protracted transit times, probably of the order of a *thousand* years for a nine light year journey. For probes into the fringes of interstellar space this might suffice, but that would be about all.

2. The fusion rocket:

Here we come up against a very difficult and fundament physical problem though it is one which in a *strictly limited sense* has already been solved. This was achieved in the early 1950s when America, Britain and Russia all produced the first so-called "hydrogen" bombs, i.e., thermonuclear explosion due to fusion. This, however, merely represents one brief and entirely uncontrollable burst of power. In this there is a parallel to the bursting of a charge of dynamite or other conventional chemical explosive. Once it has gone off it is over. It has done its stuff, for good or ill. The explosive or power generation effect will not go on repeating itself. In nature, thermonuclear power generation has been going on 93 million miles from our planet for about 5,000 million years—on the Sun. It is thermonuclear power which keeps the Sun (and all the other stars) going, the continual transition of hydrogen into helium with the attendant enormous generation of power.

So far our first feeble attempts to produce a controllable continuous thermonuclear reaction have not been noticeably successful. (H. G. Wells, probably the supreme visionary of our age, managed it in his epic novel *The World Set Free*. This resulted in unlimited power for the world. It also led to *repeating* atomic bombs—a chilling thought. But that was fiction.)

The practical difficulties are straightforward enough, and enormous. A quantity of ionized gas (i.e., plasma) would have to attain a temperature of at least 100 million degrees K. and be contained in a confined space. But what conceivable kind of material is going to restrain or confine something at such fantastically high temperatures? The only feasible method would be to trap the plasma in a supra-powerful magnetic field, a kind of magnetic tank. Such a magnetic tank would need to be free of all leaks and absolutely stable. Any instability would lead almost instantly to destruction of the "tank," loss of the plasma, and no doubt highly unpleasant consequences for anyone in the vicinity. For over thirty years or so, research along these lines has been proceeding with a noticeable lack of success—hardly very surprising given the circumstances?

Could fusion power provide a means of reliable, efficient interstellar propulsion? In one sense the position would be simplified. A rocket engine of any type, or at least its combustion chamber, is in a sense a tank with an essential and deliberate leak—the orifice or vent from which the hot gases escape thus setting up the necessary reaction or propulsion effect. Our fusion engine is then really only a magnetic tank which *must* leak at one end.

A fusion rocket engine would, experts believe, give a velocity of 10,000 km/sec. This is about 0.03C (about 6,000 miles per second). A mass ration of 20 could yield in this instance a velocity of 0.1C (18,600 miles per second). Such a craft would be capable of reaching the Alpha Centauri system in just under half a century. The time factor in relation to the human life span is still too great. There are other problems of a technical nature as well, though we will not go into them here.

3. Nuclear Pulse Rocket:

The genesis of this idea can be attributed to a German inventor, one Hermann Ganswindt, who, in 1891, hit upon the idea of a rocket propelled by a *series* of explosions. This may at a first glance seem a bit bizarre until one considers that, in essence, this is precisely the method by which a normal automobile or diesel railroad engine is propelled. Obviously we cannot apply such power units to space rockets since in the first instance they are "air breathers" and in the second the power they produce is imparted to the vehicle by means of a reciprocating engine, i.e., each separate explosion or impulse pushes down a piston which, in turn, helps to drive a crankshaft. The motor which propelled the notorious V1's or "flying bombs" developed by the Germans toward the close of World War II were in fact "air breathers" driven by the pulse principle, as their sound clearly indicated. They were of course powered by fairly conventional chemical fuels.

It is now considered that a series of explosions, damped and suitably regulated, would be capable of accelerating a rocket vehicle in a smooth fashion. Such a principle was first conceived in 1955 by Dr. Stanislaw Ulam of the Los Alamos Laboratory, New Mexico. The principle is rather crude, involving as it does the ejection of a series of atomic (i.e., fission) bombs from the rear of the space vehicle, the momentum of the explosions and debris being absorbed by a protective "pusher," or "ram" plate, secured to the spacecraft by a system of shock absorbers which would have the effect (hopefully) of imparting a relatively smooth acceleration to the craft. Unfortunately, as always, there are problems, and very real ones at that, paramount amongst them being the rather obvious one that the "pusher" plate would not for long stand up to such brutal treatment. In all likelihood the plate would quickly melt unless it were coated with a protective skin of heat ablative material, and even then it would probably not survive for long. It also soon became apparent that the shock absorbing apparatus, if it were to impart a smooth propulsion to the craft, would have to be something rather special since the pusher plate would probably weigh around several thousand tons.

Another distinctly unpleasant adjunct to such a scheme lies in the indisputable fact that fission bombs or explosions are extremely "dirty," i.e., they release very harmful reactive products. Some of this could fall to Earth if these engines were started in our "neighborhood." Moreover a massive protective shielding would be required in the interests of the crew, if they themselves were not to become the recipients of much of this very harmful radiation. And a final disincentive to the scheme is the fact that much, probably most, of the energy of each individual detonation would be lost.

A fusion (i.e., thermonuclear) reaction might not, however, be without merit in this respect. Such a reaction releases very much more energy and is also relatively "clean," i.e., it does not produce long-life harmful radioactive products. Unfortunately a thermonuclear device is detonated in the first instance by a fission detonator, the effect of which is to raise the temperature to a level in which the hydrogen to helium transition can begin. So we are back to square one, or so it certainly seems.

Freeman J. Dyson has carried out some theoretical calculations on the merits of a hypothetical fusion pulse rocket. His results are extremely interesting. The total system would have an initial mass of around 400,000 tons. Of this about 66 percent would be in the form of propellant, viz. 300,000 hydrogen bombs of one megaton yield each—rather a daunting thought—and 45,000 tons would be available as payload after making allowance for the structure of the spacecraft itself. One detonation every three seconds, he calculated, would give an acceleration of lg. Could this somehow be maintained for ten days, a final velocity of 0.03C could be attained. Thus for a nine- to ten-light-year mission 300 years would be required. Once again in terms of the human life span the transit time is totally unacceptable, though for the "generation" or "suspended animation" type of starship there *are* certain possibilities.

At this point we come to a serious refinement of the entire concept. As this was embodied in our remarks on the Daedalus Project we will not go over these points again. Daedalus may not help us much with respect to the time aspect, but the refinements to a hitherto highly dangerous and indeed rather crude concept stand out with a certain clarity. In short, Daedalus, or something very like it at present appears to lead the field.

4. The interstellar ramjet:

One of the great problems with all the propulsion systems we have so far examined has been the inescapable fact that representatives of each type *must* carry their fuel with them. The car, railroad locomotive, or ship must do the same but in these instances the ratio of fuel mass to payload is a very minor factor, relatively speaking. But even in these forms of transport a certain amount of energy is being expended merely to carry the fuel. Electric land transport in the shape of battery-operated light vehicles is at a greater disadvantage on account of the considerable weight of the lead/acid type batteries they are forced to carry. In this respect the ancient and now generally outmoded streetcar or trolley bus were at an advantage in that they drew their power from the mains by means of a pantograph or trolley in contact with a live overhead power transmission line, just as today's electric railroad locomotives do.

It would therefore be distinctly advantageous if a space ship, and especially an interstellar transport, could somehow draw its power from space itself. This seems at first very much like a pipedream, or science fiction run riot. All the systems we have examined so far have been compelled to waste much of their energy in accelerating and then transporting their fuel. An almost ideal solution to interstellar travel would be the ship which could collect its fuel requirements as it went on its way. But where is the material to come from? Space is supposedly a near perfect vacuum. Walk out from a ship onto the Moon without a space suit and evidence of that is not long in coming! The fact is however, that space is really quite far from being a total vacuum. It is filled with a very tenuous mixture of gas and dust in the form chiefly of hydrogen and helium. The density of this material is incredibly low (about 10^{-21} kgm/m^3), which works out at less than *one atom per cubic centimeter* though in the midst of the great gaseous nebulae it may be a thousand times or more greater. Space therefore is *not* a total void; material does exist there. The question now arises as to whether or not it is in a usable form for our purposes. The answer to this, it would seem, lies in the possibility of a fusion motor being able to scoop up enough of this material. In 1960, R. W. Bussard envisaged a 1,000 ton space vehicle able to maintain an acceleration of lg in interstellar space if it had an intake area in a high density region of 10,000 square kilometers and of 10 million square kilometers in one of low density. His calculations showed that the *diameter* of the intake areas would have to be of 100 and 3,000 kilometers respectively. These are clearly immense and virtually beyond the bounds of reason. It could be, however, that the intake (scoop might be a more appropriate term) could in practice be much smaller. Bussard suggested an arrangement whereby the scoop generated a very powerful magnetic field, thus effectively sucking ionized gas into the reactor system.

One adverse factor about the ramjet is the necessity for it to attain a fairly considerable velocity before the scoop arrangement is able to collect an adequate amount of material to sustain fusion reaction. This means that such an interstellar ship would have to be provided with an auxiliary propulsion system of a more conventional form in order that this critical velocity range could be attained. And this, though to a much more limited extent, brings us back to the old problem of *carrying* fuel and the associated factor of fuel mass to total mass.

Alan Bond has calculated that the ramjet becomes effective at fairly small fractions of light velocity, viz. 50 percent of total thrust at 0.02C. Such an initial boost is believed to be perfectly feasible with auxiliary conventional propulsion systems. Thereafter the ramjet would take over accelerating steadily to very much higher velocities.

At this point we come back to the old and seeming paradox of time dilation at relativistic velocities. This rather enigmatic feature has been dealt with in several other works, including *Interstellar Travel,* so there is little point in dealing with the matter at length again in these pages. Briefly, the nearer a space vessel comes to the velocity of light the shorter is the time interval to those on board compared to that for those remaining behind on Earth. At almost the velocity of light, something like 0.99C, it is reckoned that an *intergalactic* vessel could reach the nearest other galaxy to our own, two and a quarter million light years distant and return, the occupants having aged only about fifty-four years in the process. But they would return to an Earth some *four million* years older. Understandably nobody has got around to trying this out yet. Someday humanity may have to retreat from a dying Sun but it seems improbable they will look for sanctuary in another universe. It seems a long way to go.

This then is the principal relativistic effect of travel at near light velocities. With our ramjet we now come up against another. As a ramjet-propelled space vehicle approached light velocity (it could never actually attain light velocity since this must continue to be regarded as unattainable), distances in space would, by virtue of the facts outlined in the preceding paragraph, appear *shorter* to the crew. Consequently there would be an apparent increase in the quantity of elemental hydrogen inflow via the scoop. This would permit either a higher rate of acceleration to be attained, or alternatively the same acceleration as before with *reduced* engine efficiency. Thus the same acceleration could be attained in a region of space containing *less* hydrogen. Clearly the ramjet possesses an immense potential, but it is no use pretending that we are as yet anywhere near such a form of interstellar drive. Our technology is not yet up to it. Given time and a reasonable degree of fortune, it will be. But perhaps already out there in the great deeps of space, craft from more advanced technological societies from worlds of other suns have already achieved, developed, and perfected this mode of practical interstellar travel. By such means they might retreat from their own dying or rampaging sun.

5. Use of Anti-matter:

Here we have a source of energy beloved of science-fiction writers. But from the very outset it must be admitted that we are, as yet, by no means certain that anti-matter exists in practice, even if a theoretical case exists for it. The concept is based on the fact that the nucleus of an atom is *positively* charged and its orbiting electrons *negatively* charged. Anti-matter would entail a reversal of this situation, whereby the *nucleus* is *negatively* charged and the orbiting *electrons positively* so. Bring the two together and according to popular belief, an explosion of shattering immensity would be the result. But somehow control or regulate the union of the two and the power thereby derived can be put to a useful purpose, in much the same way as the power of an atom bomb is virtually the power that exists under strict and (we hope) safe conditions in nuclear generating stations.

If our universe including ourselves were constructed of anti-matter, it is doubtful if either would look or act in any way different. But in no part of the universe is anti-matter positively known to exist at the present time. It *is* true to state that small, virtually infinitesimal quantities of anti-matter particles have been created artificially—so far apparently without disastrous results.

What precisely should occur if a reasonable quantity of normal matter were brought into contact with an equivalent amount of antimatter? According to theory the two would simply annihilate one another totally, the energy thus derived being released in the form of deadly gamma radiation. Total annihilation of matter by a "mating" of two forms does not mean that nothing remains. What in effect would take place would be the 100 percent *conversion* of matter into energy. From a propulsion point of view this seems a very rosy picture indeed (for those interested in television space epics, the propulsion system of the starship U.S.S. *Enterprise* in the series "Star Trek" uses an anti-matter drive, coupled of course with the capacity to "warp" its way through space— quite a combination!).

Unfortunately, about 50 percent of the energy is released in the form of neutrinos. This form of particle radiation is one for which no effective shield is possible since it has the capacity to pass through anything. Nor can it be deflected by either magnetic or electric fields. The rest is in the form of gamma radiation, and until now all attempts to utilize these into a means of thrust have been unsuccessful. The day of the terrestrial star ship embodying an anti-matter drive is still far distant, if indeed it ever comes at all.

6. The Photon rocket:

The so-called photon rocket, could it be achieved, would simply be a space-vehicle the propulsion system of which relied in its entirety on pure electro-magnetic radiation. This has an immediate appeal for it would provide the craft with the highest exhaust velocity achievable—that of light itself. So far so good. It does not mean however that if all the ship's occupants stand in the vicinity of the propulsion tubes casting beams backward by the use of powerful search lights they are going to get very far on their way to Alpha Centauri or anywhere else. The thing would not move one millimeter though it might make a pretty sight!

The problems which arise are not greatly dissimilar to those we came up against when we considered an anti-matter drive. To attain an acceleration of just 1g, 3,000 megawatts of power per kilogram of space vehicle would be required. Such a ratio requires no further comment. The only method by which velocities representing a useful fraction of light velocity could be obtained would be by the conversion of mass to energy at a figure close to 100 percent. Only a matter/anti-matter union could produce this—so back to square one. Clearly we will not retreat from a dying sun to the stars thrusting a beam of light behind us like the tail of Halley's Comet.

These then are but pointers to what the future may hold as man first reaches out feebly for the stars. As pointers they probably have considerable validity, but it would be entirely wrong to suggest that the great star-liners of the future will either resemble them or contain power units which are merely updated and expanded versions of some of the possibilities we have been considering. A similarrity there could be, but it would probably be of the same order as that between the first transatlantic steamship and the latter-day *Queen Mary*.

If there comes a day when we on Earth and colonies on other planets and moons of the Solar System must quit the environs of the Sun for the light, warmth, and greater security of another star, and if also by then interstellar travel has become, by one means or another, safely established, then the liners and super freighters will be creations beyond our wildest imaginings—for the good and very simple reason that they will have to be! It will take more than a handful of medium-sized interstellar space transports to ensure the mass migration of an entire world, its civilization and as many of the artifacts of that civilization as possible. Its *follies* we hope will be left behind.

Which Way?

MUCH MORE POSITIVE identification of extra-solar planets and especially of their nature can only really come as a result of direct interstellar exploration. This may not lie quite so far in the distant future as was once thought, but neither will it occur next year or the year after that. Since all the tremendous difficulties inherent in interstellar travel have been dealt with fairly fully in several other books, as well as being briefly mentioned in this one, it is unnecessary to elaborate further on the matter. The problems, and they are as many as they are varied, speak loudly for themselves.

When, however, it does become a really feasible proposition, there will be no point whatsoever in heading out blindly in any direction. In that way can only lie disaster. Columbus sailed west out into the Atlantic seeking a westward route to India. Instead he discovered the Americas. He was lucky. The Americas just happened to be there. In the field of interstar voyaging such haphazard aims certainly will not suffice. The route that man is going to follow, and his specific purposes in doing so, will have to be well thought out and clearly defined beforehand.

There will be two fundamentals upon which his course of action will have to be based. These in themselves are fairly obvious and straightforward, and are as follows:

200/WHERE WILL WE GO WHEN THE SUN DIES?

a. Toward which star(s) must he set course?
b. In which direction must he travel if his object is to meet up with other galactic races?

The initial response to the first of these questions hardly calls for the use of a crystal ball. The star in question will be the nearest, our old friend Alpha Centauri which shines so beckoningly in our southern skies. Fortuitously this system contains two stars (three, if one includes the more distant Proxima Centauri—distant, that is, from the other two). One is of type G and the other of type K, both of which, especially the former, are stars fairly similar to our Sun. Moreover as we saw in our study of binary and multiple star systems the distance separating the two suns of Alpha Centauri is such that the possibility of orbiting planets is by no means impossible. To visit a star unlikely to possess planets would be totally pointless, for men and women can only make a landfall where there are planets. A trip to Sirius only eight light years distant, might from an astronomical viewpoint be highly rewarding. A close view of the famous Sirius B. (sometimes known as the "Pup"), a dense white degenerate dwarf star, could not land on Sirius A, a young, bright and intensely hot star. Neither could they land on Sirius B. The possibility of planets here is so remote as to be virtually non-existent. Perhaps some day, aeons hence, Sirius A will have its planets but by our time scale this is much too long. Barnard's Star, six light years out, might seem much more tempting. Here we very strongly suspect the existence of planets. Unfortunately Barnard's Star is a small class M red dwarf, cool by stellar standards and with therefore, a very narrow ecosphere. Planets where men could survive are not very likely in the environs of Barnard's Star. And stars lying relatively close at hand with only gas giant planets on the Jupiter/Saturn model are no use either.

In which direction then should we head after Alpha Centauri? The picture is now apparently much less clear. Perhaps after all we really *do* need that crystal ball! However let us endeavor to approach the matter in a cool and rational fashion. In so doing we will assume that no fast technique has yet been evolved, i.e., short cuts through extra-dimensional hyper-space and the like. For the moment, in thought at least, we are going to do it the hard way.

By doing so we will realize that the highway to the stars, though very broad, is also most uncomfortably long.

We must therefore assume (and it is a fairly reasonable assumption) that the direction of interstellar exploration will be toward stars with known or strongly suspected planetary familites, the initial aim being to discover which of these systems contain planets suitable for our race. Eventually our exploration of the galaxy (or at least a part of it) will constitute a slow outward expansion by the human race. As each new planetary system is colonized it will be mandatory to give much thought to selecting a new stellar destination to which mankind can expand further. With each such jump we will be getting further and further from the Sun. The decision in respect to direction would almost certainly, as before, be based on stellar evolution (i.e., type and age of star) and the likelihood of suitable planetary systems. Other factors would also come into the picture.

As we are by now well aware the Sun is merely a rather undistinguished star in a tyupical disc-shaped and very average galaxy. Its position within the galaxy, which we term the Milky Way, is neither near its center nor its outer rim. Even its position is undistinguished! In fact it lies about two-thirds of the way from the center of the galaxy, or one-third from the rim, depending upon how one cares to look at it.

The fact that the galaxy is disc shaped, fatter at the center and thinner at the rim (it has been unkindly likened to a monstrous fried egg!) is of considerable importance in our present context, for it means that the diameter/maximum thickness ratio is considerable. In fact it is about ten to one. Clearly then, we have one exploration parameter. We should travel in the *plane* of the galaxy where stellar density is greatest and not at right angles to it. Should we elect to follow the latter course for any great distance (by interstellar standards that is) we would eventually find the stars becoming fewer and fewer until eventually we passed out of the galaxy altogether into the dark, starless and terrible abyss of inter-galactic space. And if we think that the immensity between stars is tremendous, that between galaxies is infinitely more so and to an almost terrifying degree.

The reader may care to recall at this point that the *nearest* other galaxy to our own is the great spiral M31 which can just be seen by the unaided eye on a clear and moonless night in the constellation of Andromeda. We should therefore, for the most part, travel in the *plane* of the Milky Way. Here again we are confronted by a choice. Since we lie one-third from the outer perimeter of the galaxy and two-thirds from its center the correct decision is not very difficult to make. By heading towards the rim, stellar density would also begin to diminish until eventually we left the galaxy once more. We should therefore stick to the *plane* of the galaxy but preferably direct our starships toward its center. The further we travel in that direction the greater and greater would become the stellar density and consequently the greater the chances of finding suitable planets.

If we head towards the center of the galaxy at an angle of approximately 55° to the radius vector several interesting desti-

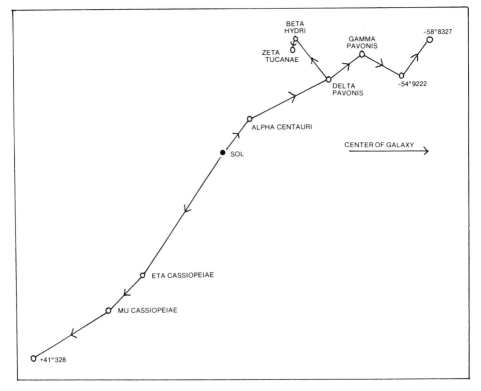

Figure 19

nations are possible. (The term radius vector is actually the line joining the focus to the body which moves about it in an elliptic orbit.) In the case of the Solar System it is the imaginary line from the Sun to any one of the planets. In our present context it is the imaginary line between the Sun and the center of the galaxy (Figure 19). The first objective is, hardly surprisingly, Alpha Centauri. From there, or at least from a suitable planet, starships could head for the star Delta Pavonis which lies in the southern constellation of Pavo. This is also a G type star similar in most respects to the Sun. Appropriate planets orbiting this star could lead to yet another growing and maturing terrestrial colony. Starships from the Delta Pavonis system would find their next most suitable objective in another of the stars in Pavo, this time Gamma Pavonis. This is a class F star and therefore slightly younger and hotter than our Sun, yet old enough to have acquired planets. Because of its higher temperature we could confidently expect a wider ecosphere in which suitable planets might orbit. Terrestrial civilization (if by then it could strictly be termed "terrestrial"), still heading outwards, would probably favor, as their next objective, a binary system unnamed, but known to astronomers as -54° 9222 and thence in time from there to another G type star, -58° 8327. These stars lie in the southern constellations, Vela and Carina respectively.

Like a railroad track, there is a point on the main line where a branch diverges. On our great interstellar railroad in the skies that branch occurs at Delta Pavonis. Our distant descendants could, at this point, continue either along the main line to Gamma Pavonis and two other stars beyond *or* they could take the branch which would lead to the star Beta Hydri and, a civilization once established there, to Beta Tucanae. Both of these are class G stars similar to the Sun.

But now let us go back to the beginning and to our Sun. A few paragraphs back we said that travel toward the rim of the galaxy rather than towards its center would become steadily more counter productive on account of diminishing stellar density. In its essentials this fact remains perfectly true. We could however still expand in that direction, though only to a more limited extent. Let us therefore on this occasion head not for our old friend and "close" neighbor Alpha Centauri; starships, we will assume, have already gone forth in that direction. Our second wave of would-be colonizers heads this time in a roughly opposite direction (*away* from the galactic center) at a radius vector of 50°

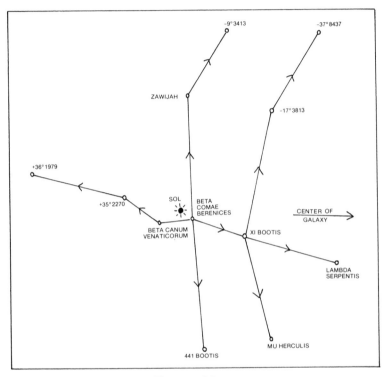

Figure 20

(Figure 20) toward the northern constellation star Eta Cassiopeiae. This is also a binary system, the two stars comprising it being of class G and M respectively. The G star in this case is *identical* to our Sun. The class M star however would be a different proposition. With a surface temperature in the region of 4,000°K it would be too cool for suitable planets and would probably be ignored. A new terrestrial civilization once established there might then look toward the star Mu Cassiopeiae. This is also a binary system, one of the components again being very similar to our Sun. From there, in the fullness of time, a new colonizing wave could head for the star known as +41° 328. This lies in the northern constellation of Perseus and is also a near-solar type.

The original colonizing wave going via Alpha Centauri to -58° 8327 would in the end have carried terrestrial civilization a distance of 47 light years from the Sun, the second only 38. We can see therefore that the first wave, taking in the nearest star Alpha Centauri, spreads terrestrial civilization the greater distance. Combinging the two we find that mankind would have spread his civilization over a distance of some 85 light years. We should also note however that this is more or less a "straight line" of stellar colonization. The idea of a "shell" of creeping, expanding civilization, ours or anyone else's, sounds fine but in practice the thing is more likely to resolve itself into a system of expanding arms or "tentacles."

We could also, again to a more limited degree, travel to the north and to the south of the galactic plane. What are our options here? Let us start with routes to the north. First destination might well be the star Beta Comae Berenices, second brightest star in the rather undistinguished northern constellation of Coma Berenices, which lies between the stars of Leo and those of Boötes, and thence in turn to XI Bootis. These migrations are, in fact much more toward the *center* of the galaxy but XI Bootis represents a good point in space for further spread of the terrestrial tentacle away from the plane of the galaxy. From XI Bootis colonizing expeditions could head in turn toward the stars -17° 3813 and -37° 8437. These stars are in the constellations Eridanus and Puppis respectively. Alternatively from XI Bootis starships could head in the generally opposite direction toward the star Mu Herculis. The civilizations established on a planet or planets of Beta Comae Berenices could, as well, (or instead of) send ships towards the star known as Zawijah (Beta Virginis) and thence in time and in turn toward -9° 3413 in the constellation Eridanus. The opposite direction from Beta Coronae Berenices would lead to the binary system 44 Bootis.

To the south of the galactic plane there are also several options open. From the Sun we could first proceed in the plane of the galaxy *away* from the center toward the star Tau Ceti. This would represent a particularly good choice since this star has long been favored in the planetary sense and was indeed one of the two stars chosen for the now almost legendary Project Ozma inspired by Dr. Frank Drake back in 1960. It lies only 10.2 light years distant and is of class K. It is thus slightly older and cooler than the Sun. From Tau Ceti, assuming a suitable planet or planets existed there, a farther colonizing wave might consider the star 82 Eridani at virtually 90° to the plane of the galaxy. Were conditions favorable for colonization there, the next move might be one toward Gamma Leporis in the constellation Lupus. This involves travelling toward the galactic rim at approximately 40° to the galactic plane. Alternatively, by heading at almost 90° in the opposite direction from Tau Ceti, star ships could eventually reach Van Maanen's Star, a further two and a half light years distant. This is a class F star, slightly younger and hotter than our Sun and perhaps therefore again with a broader ecosphere.

All the foregoing represent a mere skeleton of the possibilities. To the reader with a conventional star atlas the suggested routes might seem as if we were moving about the heavens in the most haphazard of ways. What he or she must remember however, is the fact that by looking at a star atlas or standing out under the night sky the heavens are being seen as a "dome." *Relative* distances are *not* apparent. Some stars are brighter than others. This may mean they are nearer. On the other hand some are very bright yet much further away than certain fainter ones. Their real or intrinsic brightness just happens to be greater. Deneb in the constellation of Cygnus the Swan has a luminosity at least 10,000 times that of our Sun, and lies over 600 light years distant. Sirius, which appears to us as the brightest star in our skies, is only 26 times as bright as the Sun. However *it* lies only eight light years away! Proxima Centauri, third partner in the Alpha Centauri system is the nearest star to us, at a distance of 4.2 light years. Its absolute luminosity is only 0.0001 that of the Sun which renders it totally invisible to the naked eye.

As a further reminder of the true position consider the magnificent and well-known constellation of Orion, the Hunter. Its configuration seen from a point much further out in space would render it totally different. From Betelgeuse in Orion to Aldebaran in Taurus the Bull seems, from Earth, not too far. but Betelgeuse lies 190 light years from the Sun, Aldebaran only 57!

It also cannot be stressed too strongly that all this talk of routes does *not,* and never could, represent the itinerary for a conducted, package tour of the stars by a single generation. What we are contemplating is the slow expansion of our race as a consequence of interstellar exploration and colonization: Earthmen and women who on a planet or planets of Alpha Centauri create a new advanced technological civilization. Representatives and colonists from this then set out to repeat the process on an appropriate planet or planets of Delta Pavonis. Whether or not suitable planets will be found in orbit around each destination star is impossible to say. In many cases there is certainty of disappointment—and perhaps as a consequence, tragedy. An expanding but still basically terrestrial civilization will then have to mourn, and re-think.

With a more highly advanced technology than we have at present our distant descendants will be in a much better position to make the leap with correspondingly less risk. High speed probes fired ahead by the oncoming star ships might just conceivably be able to provide and send back the vital information concerning the destination star. As a consequence those in charge of the star-ship fleet would be in a better position to know whether to proceed or turn around and go back while there was yet time.

Now we come briefly to the second of the fundamental questions we set outselves at the commencement of this chapter, i.e., in which direction must mankind travel if his object is to meet up with other intelligent species in the galaxy? Presumably if he wishes to meet up with alien beings not unlike ourselves, or at least caricatures of our own kind, he will be looking for planets like our own orbiting suns like our own. In this case he will, to all intents and purposes, be doing something along the lines already specified. by then of course, radio signals from certain extra-solar planets may have been picked up by radio telescopes on Earth, in which case he will know precisely where to head.

But here a vital and highly important point arises. Would terrestrial civilization bent on creating its own hegemony over a portion of the galaxy want to find beings on planets of which it wishes to take absolute possession and control? Very likely it would not, a viewpoint shared equally by the inhabitants of these planets. If they happened to be sufficiently advanced technologically, open war would almost certainly result with the men of Earth clearly and unmistakeably the aggressors. With a backward race the prospects could be very different. Here would be a direct parallel with the policy of empire building carried out on Earth over the centuries and presently the cause of so much trouble, bloodshed, and international tension. Subject peoples, despite benevolence by their rulers, do not appreciate being subject forever.

The trouble is that on planets like our own, orbiting suns of an identical or very similar nature, biology may easily have, probably has, followed a roughly parallel course. Thus in many cases inhabitants of one sort or another will probably exist. This is an impasse around which it is difficult to find a way. Planets devoid of life entirely, both animal and vegetable or appropriate only to life forms vastly different from our own would probably be of no use to us, in which case the problem does not arise (e.g., a planet with an atmosphere essentially of chlorine or one of pure nitrogen).

We spoke earlier of the eventual creeping tentacles of terrestrial civilization spreading out among the nearer stars, and of branches from these tentacles eventually following suit. If other intelligent technological civilizations exist within the galaxy, especially among the local cluster of stars of which our Sun is one, there is a very strong possibility that sooner or later one or more of these terrestrial tentacles is going to meet one of an alien kind. The results should be interesting.

Cities in Space

ONE POSSIBILITY WE have so far neither examined nor considered is that of colonies *in* space (interplanetary or interstellar) living on what can only be described as artificial planets. The concept is assuredly not a new one, and science-fiction writers have been making extensive use of it for years.

First though let us see how the idea fits in with the title of this book, if we choose to take the title in its most literal sense. Let us assume (and this still remains a very considerable assumption) that our benevolent and seemingly so stable Sun, contrary to all the postulates and deeply held beliefs of astrophysicists is, after all, due to go "nova" in the foreseeable future. Let us further assume that this nova is going to be no minor affair. To escape its cataclysmic effects, mankind must quit not the Earth and the immediate environs of the Sun but the entire Solar System. And to complete the scenario, although he is by now aware that several of the nearer stars have suitable planets, he still lacks the *means* of getting there. In these circumstances all he can do is to utilize his most advanced technology in saving a small portion of the human race by the creation of artificial planetoids, giant space stations, veritable cities in space. This represents the most extreme set of circumstances possible to which we will return later in the chapter. For the present it might be of interest to examine and consider a rather less exacting scenario but one which, unlike the nova, could and probably will occur—one that is probably occurring already. Let us look very closely at this particular scenario.

Recently, thanks to interplanetary probes, it has become all too obvious that none of the planets or moons in the Solar System are capable of sustaining human life. It might in time be possible to alter dramatically the climate of Venus to our advantage. Small colonies might be created on Mars and the Moon but only under highly artificial conditions. And, of course, so far as the present or foreseeable future is concerned there is simply nowhere else to go.

Now at the present time two highly dangerous tendencies are already rearing their ugly heads—overpopulation of our planet and a world-wide energy shortage. The cure for over-population lies in our own hands if we have the sense to appreciate the danger and take the obvious necessary steps. This is a socio-economic-biological problem which hardly belongs to these pages. The other is much more urgent.

Few of us today have not heard of or felt the effects of the gathering energy crisis. For several generations now we have used (and to a large extent wasted) our invaluable stocks of fossil fuels, which are only replaceable on the geological time scale in which a million years is as a day. The shortage is particularly acute in the case of oil, and already we see it reflected in the cost of heating oil for our homes and gasoline for our automobiles. The price aspect is, of course, presently being aggravated by the greed and blackmailing tactics of many of the nations in whose territories nature happened to place vast oil reserves. Coal is being dragged out of the ground at a prodigious and ever-increasing rate. Here the reserves are much greater though still, it must be emphasized, not limitless. And of course, the more oil and coal we burn as fuel, the more carbon dioxide we pump into our atmosphere. We run an increasing risk of setting in motion the celebrated "greenhouse effect" which could partially melt our pole caps, raise substantially the level of our oceans, and as a consequence inundate vast tracts of low-lying territory adjoining coastal regions.

Nuclear power provides an answer of sorts, but in view of the dangers apparently inherent in its use this has dropped greatly in the popularity polls over the past decade and especially since the regrettable incident at Three Mile Island, Pennsylvania. It may well be that the dangers are overstated. Nevertheless this is a relatively recent technology, many long-lasting and inherently dangerous radioactive residues are produced, and it is obvious that in this field of human endeavor we should probably make haste slowly (and thoughtfully).

So where does all this leave us? There is geothermal energy and in certain volcanic regions, notably Italy, Iceland and New Zealand, this has already proved and is proving of considerable worth. But in the light of the world's escalating energy requirements it is merely a drop in the bucket. There is also wind power which may yet prove of inestimable value. There is, too, the great and unending supply of energy manifested by the waves of ocean and sea. Unfortunately the tapping of this source produces considerable technical problems and it must be admitted that most of the schemes produced so far seem strangely unreal and generally impracticable.

It so happens, however, that we have a very convenient and extremely large and efficient nuclear pile working for us only ninety-odd million miles distant; one which is going to do its stuff for a few thousand million years yet. We refer, of course, to the Sun which endlessly drenches our planet in a great flood of energy. The problem is how to collect, harness, and put that power to practical use. So far most existing schemes have been largely experimental, and either not very successful or not particularly efficient. The problems speak for themselves. In parts of the world where the hours of sunlight are many and the heat of the Sun is very strong, large areas are still generally required to collect, reflect, and focus the heat in order to transform it into readily usable forms of energy. In countries where the Sun is only rarely seen and gives little warmth the problem of benefiting from the Sun's energy is much more difficult.

Back in 1959, during the early dawn of the Space Age, in a paper entitled "Search for Artificial Stellar Souces of Infra-red Radiation," F. J. Dyson made the point that, even allowing for modest growth rates in energy consumption, the human race would need to consume the entire solar energy flux bathing this planet on a timescale more or less comparable with the present length of written history. But for years and long before 1959 our growth rates in energy consumption have been anything but modest. Indeed they are becoming exponential. Tomorrow, it would seem, simply doesn't matter to most of us.

We accept then that despite the great liberality of the Sun in respect to energy, the difficulties in collecting and using it are quite considerable. At the present time it is rather like trying to collect water in a sieve instead of in a bucket. But if part of the human race were to move into space *and stay there* energy flux from the Sun could be used directly and with relative ease. The idea is not original by any means and was first suggested by Tsiolkovsky as far back as 1912. Later it was reactivated by J. D. Bernal in 1929 and by Arthur C. Clarke in 1951. It was left to G. K. O'Neill, however, to place the idea firmly on the map. Whilst the others had expressed the concept in fairly vague, generalized terms O'Neill began to extrapolate to a considerable degree. What he envisaged was a very large space structure, the interior of which could be made to resemble the natural environment of the Earth's surface. One of the principal features in this respect was to provide the illusion of the Sun's apparent daily path across the sky. This, he reckoned, could be achieved by the use of three hinged mirrors placed around the axis of the cylindrical vessel at intervals of 120°. By swinging these mirrors through 90° it would be possible, he maintained, to reflect sunflight into the space city through an angle of 180°. It must be conceded however that the strength of current materials are insufficient to achieve this on the scale suggested by O'Neill. Nevertheless it is the future we are looking toward when materials fully capable of providing an arrangement of the necessary strength are an accomplished fact. A great orbiting city would, therefore, be able to use the tremendous force of solar energy bathing the Earth and its environs. Batteries of strategically placed photo-electric cells would supply limitless amounts of cheap, clean electrical energy.

The escalating birth-rate on Earth can only be countered by having fewer people born. For socio-biological reasons this seems highly improbable. Mankind will thus expand to fill all the living space available. In *Interstellar Travel* I quoted the facts and figures relating to population increase and found they could only be described as frightening. Living space on Earth is strictly limited (as are its resources), therefore new living space must be found.

This could be on (or even beneath) the surfaces of certain planets and moons within the Solar System, and though such extra-terrestrial colonies might absorb a proportion of our planet's surplus population, further measures would obviously be necessary. In this respect artificial construction in space would assist, though this is certainly not to suggest that a number of space cities orbiting at various points in the inner Solar System would solve the population explosion problem. At best they would merely provide a further palliative.

We have then, in imagination, provided our space city with limitless energy and the pleasing effect of a pseudo-Sun. One other essential can never be overlooked, that of gravity. In other words a pseudo-gravity must somehow be provided as well. In science fiction this represents no problem as a rule. Someone merely switches it on! Here we must try to come up with something a little better. So far as we now know the only method whereby this could reasonably be achieved would be by reaction to the centripetal force of a rotating body. In other words the vast toroid would have to rotate and centrifugal force, which is opposite and equal to centripetal, would then provide the gravitational effect.

When we talk of "space cities" however we are not merely thinking in terms of a vast space craft. Irrespective of how large this happened to be, the feeling of eternal confinement within metal walls would still be present, though obviously it would affect some people more than others. In such circumstances feelings of claustrophobia could easily become rampant. Claustrophobia is generally associated with very closely confined spaces. Aboard one of today's wide-bodied jets at 35,000 feet we know that the environment outside is hostile and would quickly and effectively kill us. So we sit comfortably warm and contented, thankful to be within the embracing security of the airplane's metal shell. But if by some peculiar freak of chance we had to spend weeks or months within these metal walls feelings of claustrophobia would almost certainly set in.

Our space city should constitute a habitat capable of accommodating about three million people in comfort and security. The essential requirements would be considerable. These would have to include the provision of a pleasant and temperate climate, a terrestrial-type oxygen/nitrogen atmosphere at normal sea level pressure and an acceptance degree of humidity. The city would have to be shielded from all forms of harmful radiation and, as already outlined, possess the means of creating the illusion of a terrestrial night and day cycle. We would also want some green fields, suburbs and hills.

The population would, of course, have to be fed and to produce the necessary food by virtue of its own labors and resources. Thus a large agricultural area would be necessary. In the case of very small space settlements, hydroponics, the technique of growing plants over water containing suitable chemicals rather than in soil, might just suffice. Such a method would be totally inadequate in the type of space city we are presently envisaging, though it could constitute a useful and most welcome adjunct. It has been suggested that the area required for agricultural produce would be on the order of fifty square meters per person and the minimum *total* area required for a population of three million people about two hundred square kilometers.

A space city in its design must be one to preclude any possibility of claustrophobia. It must therefore be very large and resemble the surface of our planet. What has recently been proposed is a hollow cylindrical shell 16 kilometers (10 miles) in diameter, 16 kilometers (10 miles) in length, with semi-toroidal edges 1.6 kilometers (1 mile) in diameter. (Figures 21 and 22.) A recent excellent paper by T. Hassall in *Spaceflight* (21, 12, pp. 504-511) suggests a central city having an area of 16 square kilometers with living space for 1.2 million people, the remaining population being housed in two linear cities about fifty kilometers in length. There would be three time zones each eight hours apart. In each would lie a central city.

Space cities built along these lines would depend for much of their effect on the synthetic "Sun" apparently crossing the sky as mentioned earlier. An imitation Moon showing its phases might even be technically feasible. The fields and hills and similar landscaping, though composed of real soil and real rock, would be equally synthetic as would the clouds (produced by virtue of a shutter arrangement cutting out some of the Sun's incoming heat, thereby producing the necessary degree of condensation). Nevertheless the impression would still be a natural one, though the magnitude of the containing vessel, whilst it might allow also for small lakes or meandering streams among the hills, could not be expected to allow for seas and rivers. "Artificial" rain from the artificially-generated clouds might be a possibility. These would replenish the lakes while evaporation from the lakes would assist in cloud generation.

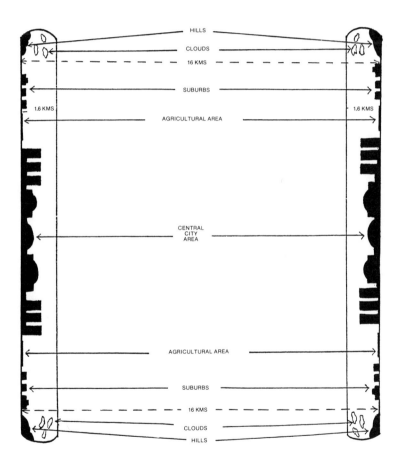

SPACE CITY

Figure 21

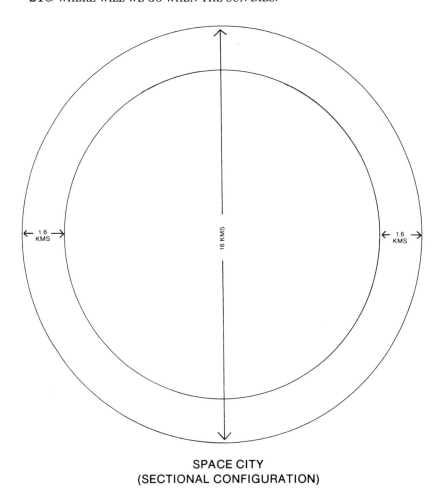

SPACE CITY
(SECTIONAL CONFIGURATION)

Figure 22

At the present point in human history such vast artificial systems orbiting in space may seem very much like science fiction—but then so fifty years ago did the prospect of men on the Moon and televised pictures from Mars. Surrounded by the airless and utterly hostile environment of space we could have the dream city of the future, plus fields, hills, clouds and perhaps even a lake or two. Absurd? But would not the plan of a Boeing 747 or a Douglas DC10 have seemed equally absurd to the Wright brothers and those other worthy pioneers of the aeronautical world? It cannot be sufficiently stressed however, that such orbiting communities could not be expected to provide an answer to the population explosion already taking place on our world. Even if space cities could be stepped up to take ten million people each and there were a hundred such "cities" this would only account for a thousand million people. This may seem a lot. Whether it is or not depends largely on the criteria being employed. So far as the population explosion is concerned it represents an infinitesimal fraction.

Another aspect which clearly would require to be taken into account are the criteria necessary to assist in the selection of citizens for these orbiting Utopias. In the first instance these could only be chosen from those who *wanted* to go. No doubt however there would be plenty of volunteers, many people anxious to quite literally "get away from it all!" Those in charge of the project would then have some rather difficult decisions to make. If a particular orbiting space city is designed to accommodate three million people then the question of over-population must be carefully considered. If the city is populated by a certain number of youngish men and an equal number of youngish women then quite soon there are going to be a lot of babies around and the population could shoot up to unacceptable levels. Since the death rate will be low because of the youth of the occupants, this cannot compensate. It would then seem essential to populate the space city with a proportion of older people and relatively fewer *young* men and women. This with reasonable luck should lead to a balance. If these fewer younger people still had more babies than was first anticipated, then the old problem would remain.

The question could of course be looked at in another way. The circumstances of the situation and its semi-idyllic nature might tempt young women to have fewer babies and therefore lessened responsibilities so as to enjoy more freedom. Dangerous under-population could then result. However this is the more unlikely eventuality. The maternal instinct would probably remain strong in young women and even if under-population began to result, more people could be brought "up from Earth." The remedy for over-population might seem to be equally simple. This probably would not be so. On what basis would the choice of those expelled "back to Earth" be decided? This could be a tricky one. The wisest course would probably be to under-populate in the first instance, give say to a space city designed to accommodate three million souls a mere two million but making the bulk of this population youngish men and women. The main aim must be to achieve balance.

Another problem would also have to be considered. When people die something has to be done with their mortal remains. Here on Earth this is generally achieved by burial or cremation and, to a limited extent, by interment at sea. None of these options are open in an orbiting space city. The easy and obvious option could hardly be invoked, i.e., casting the corpses into space through specially designed air-lock shutes. Space, we must remember, is a vacuum. A corpse would therefore burst messily! This would hardly be a comfort to relatives of the deceased, nor does it seem either pleasant or desirable to have an increasing amount of human remains in pieces orbiting with and just outside the ship. So what *is* to be done? In a much earlier book, *Journey to Alpha Centauri* (1965), I suggested that corpses might be placed in pressurized "shrouds" and shot off into deep space. Presumably this would be expensive. The thought of eventually having one's remains drift eternally among the stars seems rather pleasant in comparison to burial in the ground, at sea, or by cremation.

Space cities could be designed either to orbit Earth or to be placed in an independent orbit around the Sun, say slightly to the sunward side of Earth's orbit. In the first instance Earth would always be close in the event of anything going seriously amiss; not so in the second where at times the essential mechanics of the Solar System would place Earth and space city in conjunction, i.e., on opposite sides of the Sun and therefore extremely remote from one another.

Now let us consider the first scenario which we mentioned briefly at the beginning of this chapter. This is a tougher one because in this instance we are assuming that retreat from the entire Solar System has become mandatory, though the means to reach suitable planets of other stars has *not* been attained. In these circumstances everything is going to be different—very different! For a start this space will not be an orbiting one but merely a travelling one headed in space for nowhere in particular.

No longer will an Earth be at hand to offer succor. What could be most likely is a colony of these things with the means of brief propulsion periods so that all could remain close together.

All this would really constitute a last desperate attempt by a relatively few people to escape the death of the Solar System but it would probably represent more a stay of execution than a reprieve. The problems which would beset such a colony of space cities are legion. For a start, how would the colonists be chosen? We must assume that any circumstances necessitating a quitting of the entire Solar System would be accompanied by anarchy on Earth. The law of the jungle would prevail. The instinct for self-preservation is an intensely strong one, and there would be few who would not want or feel they had an equal right to escape the coming cataclysm whatever its nature.

The next problem would be one of selection. If such a space colony were to survive for any period of time it would require the presence of as many of the world's best administrators, scientists, technologists and doctors as possible. There could be no doubt that such people must have priority. The question of age and sex would also have to be considered closely. If this were important in an orbiting space city it is infinitely more so now since there would be no Earth to take surplus population or supply numbers to make up a declining one.

Another great difficulty would be the fact that the Sun would be absent. No system of mirrors to simulate hours of daylight and darkness would avail now. Lighting, heating, and power would have to come from an indigenous source which could surely only be nuclear (or perhaps thermonuclear) in character. Use might be made of interstellar hydrogen as outlined in chapter 12.

Despite the most careful planning, the most lavish appointments and equipment, it is hard to see such colonies surviving indefinitely unless our technology had advanced very much further from the point at which it rests today. Such a scheme however, could easily merge into a plan to reach suitable worlds of other stars by means of "generation travel." No doubt by now most readers are familiar with this concept which involves the dispatch of a vessel or vessels from the Solar System toward a "nearby" star (always a very relative term) considered to possess a suitable planet (or planets). The personnel are carefully chosen and genetically suitable, young men and women, a very high proportion of them skilled in the arts, crafts, and sciences. Several generations are born and die before the great starships reach their chosen havens, so that it is the descendants of the original occupants who land on the destination planet. Any such craft, heading for the nearest star, Alpha Centauri (4.3 light years distant) at a velocity of 0.02C (3,720 miles per second) would require slightly over two hundred years to make the journey. In this case it would be the *eighth* generation that would eventually reach the chosen planet. We would trust it would be to their liking when they got there!

Writing of even longer "generation" journeys to the stars in the journal of the British Interplanetary Society in 1952, Dr. L. R. Shepherd wrote as follows: "It would be as though the vessel had set out for its destination under the command of King Canute (died A.D. 1035) and arrived with President Truman in control (a 910 year voyage). The original crew would be legendary figures in the minds of those who finally came to the new world. Between them would lie the drama of perhaps 10,000 souls who had been born and had lived and died in an alien world without ever knowing a natural home."

The prospect of "cryogenic" travel (i.e., deep freezing the occupants of the ships) must also be given consideration. This idea is movingly and magnificently portrayed by noted science-fiction writer Arthur C. Clarke in his short story "Songs of Distant Earth." In this excellent narrative the starship *Magellan* is, for technical reasons, compelled to call at a planet of another star already populated by descendants of people from Earth. The *Magellan* is a "cryogenic" ship in which the occupants lie in death-like sleep from which they will only be aroused a century or so hence when the ship reaches a planet of its destination star. In the emergency only a few key personnel are "awakened" and are shuttled down to the planet by small craft. The poignancy of the whole business is well brought out by the author. On the planet by which *Magellan* has halted, a male crew member and a girl of that planet meet and fall in love. She wants him to stay and let *Magellan* proceed on its way without him. Sadly he tells her this cannot be. To prove why, he has her shuttled up to *Magellan* where he shows her a girl lying in a trance-like sleep. This girl is his wife! Sadly the other girl returns to the surface of the planet and the mighty *Magellan* resumes its journey. She realizes that long after she is dead and forgotten the man she loves will be resuming life on a new world with his wife. For his part the man realizes that when he does, the girl who has come to mean so much to him will have been dead for the better part of a century!

If interstellar travel is ever to be without the poignances of "generation" and "cryogenic" travel, then techniques will have to be developed that can telescope time and distance. These were fairly fully covered in *Interstellar Travel: Past, Present and Future,* to which the interested reader is referred. Only by such techniques could that part of science fiction come true in which scheduled interstellar space lines ply the galaxy, or at least parts of it, in a monstrous extrapolation of conventional air-line travel today. If we are ever to have this, the great vessels, their passengers and colonists, their daring commanders, their pretty stewardesses and all the rest, then the ships must surely ply another dimension or at least find some convenient bypass "tunnel" in space. If these things are not possible then journeys to the stars can only be done the hard way, for these will be journeys from which there can be no return.

Quid Pro Quo

IN CONSIDERING POSSIBLE planets of other stars, we have always
had in the back of our minds (as the title of this book indicates)
other appropriate habitats to which we, or at least our distant
descendants, could depart should our Sun ever "go nova," start to
emit lethal levels of flare radiation, or simply cool down or heat up
to an unacceptable level. Present astrophysical reasoning is that,
while these things can and do happen to certain other stars, they
will *not* occur in the case of our Sun. What seems much more
probable is that the Sun, round about 5,000 million years from
now, will swell out to become a bloated red giant star consuming
in the process all the inner planets including Earth. How accurate
may be the estimate of 5,000 million years is probably open to
doubt, but at least the time scale is almost certainly something of
that order.

This calamitous episode will not mark the end of the Sun as a
star but it will assuredly indicate the beginning of that end, for
this phase is indicative of a star's rapidly increasing senility (e.g.,
Betelguese, Antares, Aldebaran to name but three). When it
eventually contracts and condenses into a fantastically dense
white dwarf star its demise, on the astronomical time scale, is not
far distant. Such is the life cycle of a star—from cloud of elemental hydrogen gas to a dark, cold and forever inert globe. Even the

eternal stars are *not* after all eternal. Neither in all probability is the universe itself, though the time scale in this case is tremendous beyond imagining.

Now 5,000 million years ago our essential life-giving luminary came into existence, and probably after another 5,000 million years its death throes will begin. This makes our Sun a truly mature and middle-aged star. When it came into existence from its primordial hydrogen cloud so too, or perhaps very shortly after, (astronomically speaking) did Earth and its sister planets. Earth, in its present form, and especially the life that grows, walks, crawls, slithers, swims, and flies about its surface, are all extremely recent on the scale of geological time. So in 5,000 million years Earth and its inhabitants may have changed so drastically and dramatically that where to migrate when the Sun dies could perhaps only represent an academic question. Will Earth, as a pleasant, habitable planet last the merest fraction of that time? The manner in which some of its supposedly mature and intelligent occupants presently conduct themselves, and the way in which they treat their planet and its resources renders this a very debatable point indeed. Nevertheless we must endeavor to remain optimistic about this, though at times this is extremely difficult. During the next thousand years we should have colonized the Solar System, or at least those parts of it which are colonizable. New technologies should help us do this. Right now human beings could not exist on Mars but in time, and in limited numbers, there seems no reason why they should not.

However, removing part of Earth's teeming population to other planets and moons in the Solar System avails us nothing if the Sun, by virtue of some disastrous and unexpected action, renders our system untenable. True salvation, the only possible salvation in such disastrous circumstances, is transtellar migration to a world or worlds of a kindlier and more predictable sun.

This brings us automatically to the crux of the present chapter. Has something like this had to be done already by advanced cosmic communities in this and in other galaxies; and, even more pertinently, could *our* Sun represent possible salvation to a highly technological and sophisticated society whose own sun has rendered *their* system untenable? Should this ever happen then it could easily be a case of God help humanity! However, the attack by aliens theme has been dealt with already in a number of works, including my own book, *Space Weapons/Space War*, so there is no point in returning to it here. Neither have we the remotest intention of contributing further to the highly vexed and somewhat emotive subject of U.F.O.s, "flying saucers," and other peculiar forms of space-borne crockery. The question of these objects remains a very open one.

Indeed the entire matter of extra-terrestrial and extra-solar life remains fairly controversial. A few decades ago we here on Earth regarded ourselves as unique, alone in the universe, potential lords of all we surveyed. Very gradually belief began to swing perceptibly in the other direction. Over the past few years this trend has escalated to such an extent that it may be that it has now gone a little too far.

Recently there has been in evidence a certain polarization in the situation, with one small camp again reverting to the alone-in-the-universe theme, but a very much larger one still adhering to the other viewpoint. Which is right? This remains the perennial question. The "alone" or "almost alone" camp is represented to a large extent by Dr. Michael H. Hart of Trinity University in San Antonio, Texas, who bases his case very largely on a computer analysis of hypothetical planets, introducing those features that would seem essential in order to produce advanced civilizations like our own. As a consequence he believes that, far from being common, such life must be exceedingly rare in the universe, perhaps even to the extent of our life here on Earth being unique. He says, "exobiologists and other believers in advanced extra-terrestrial civilizations have estimated the number of these civilizations at anything between 50,000 to one billion or more. This study shows that such theoretical estimates are a hundred to a thousand times too high."

His assumptions (and it must be admitted these are eminently reasonable) are that prevailing temperatures must be moderate and that these must remain so continuously for at least 3.7 billion years, this being the period which elapsed on this planet between the origin of life and our present level of evolution. Dr. Hart concludes that our planet only just made it as a cradle for life. "At a distance of 93 million miles from the Sun" he says, "terrestrial temperatures have supported life. But if our Earth had been slung into an orbit only five percent closer to the Sun a run-away 'green house' effect would have turned the planet into something like Venus—a cloud-shrouded planet with temperatures close to 900°F. If, on the other hand, we had only been one percent farther from the Sun when the Earth came into being, runaway glaciation would have enveloped the Earth completely, and 1.7 billion years ago our planet would have become a barren, frigid desert similar to Mars." In other words a mere 6 percent has resulted in our presence on this planet: 4.7 million miles closer to the Sun conditions would have been too hot, with all the inevitable consequences that heat brings; one million miles farther from the Sun, conditions would have been too cold. All in all a pretty close run thing it would seem!

These points, the reader will recall, were briefly pointed out in an earlier chapter. All this of course assumes that life's initiation and the many forms it assumes must be more or less carbon copies of our own (ironically the term "carbon" in this instance possesses connotations which are real as well as metaphorical). However, to extrapolate using other elements in place of carbon, hydrogen, oxygen, and nitrogen inevitably introduces a very high degree of speculation. There may be forms of life based on other elements, as we discussed briefly in chapter 11. Here on Earth we have no evidence or experience of them, and therefore in my opinion it is interesting but inadvisable to pursue this line of reasoning too far. In science fiction the practice is acceptable but in science we must stick to facts as they are, not as they *might* be (or even as we might like them to be!).

Nevertheless, accepting that Hart is correct in his assumptions (and there is little doubt he is), the fact remains that, so great is the number of stars in our galaxy, duplication and near duplication of terrestrial conditions *must* have occurred in a very great number of instances. This, of course, Dr. Hart is admitting when he remarks that present estimates could merely be a hundred to a thousand times too high.

Dr. Hart's contention that we might conceivably be *entirely alone* in the universe still remains a minority one among other scientists. The opposition is headed by the redoubtable Dr. Carl Sagan, though even he cautions against undue optimism in the search for other cosmic communities. Recently he wrote, "Some scientists working on the question of extra-terrestrial intelligence, myself among them, have attempted to estimate the number of advanced technical civilizations in the Milky Way galaxy—that is, societies capable of radio-astronomy. Such estimates are little better than guesses, though I believe that of the 250 billion stars in our galaxy there are about a million with planets supporting technical civilization."

Of Hart's premise Sagan says, "I believe he (Hart) sells short the adaptability of life to the hostile conditions that may pertain in any given corner of the universe. I also believe he neglects the evolutionary process that molds our particular life forms to fit the conditions of our Earth. He assumes that the Earth just *chanced* (author's italics) to be the way our life forms require it to be."

If then other civilizations with the capacity for instellar travel exist there could be some who, by virtue of a dying or a potentially lethal sun, have been forced to evacuate their planetary systems to seek, and hopefully find, a locale for the continuation of their species on the planet or planets of adopted suns. Could *our* Sun ever satisfy this requirement so that, instead of Homo sapiens quitting the Solar System, we find ourselves the unwilling hosts to other less fortunate and perhaps even highly ruthless beings from a star a few light years distant? Since this was the essential theme of my book *Space Weapons/Space War* it is not one we will attempt to pursue here. Is it possible however, that the attempt to reach our system *was* once made but, apart from the arrival of a few advanced craft and their occupants, it proved abortive and was subsequently abandoned? Perhaps at close range Earth looked less appropriate.

Despite the odd claims of many writers and researchers in this field, there is as yet *no* really valid, unequivocal evidence of alien artifacts or presence here on Earth. (There *is* evidence of plenty of wishful thinking on the subject however.) Could they have reached Mars? It is too early yet to say. What of the Moon, however? So far twelve American astronauts have briefly walked upon and explored small areas of its stark and inhospitable surface. Nothing that could be remotely construed as alien artifacts or evidence of alien presence has been found. The Moon has been orbited and photographed at close range as well as being scrutinized for many years by astronomers using giant optical telescopes here on Earth. No such evidence has been detected—*or has it?*

Fairly recently certain unique details *have* been observed on the Moon. These just *might* come into the category of alien artifacts or evidence of an erstwhile brief alien presence. (The word "might" in this instance cannot be emphasized too strongly!) Let us look briefly at the evidence, if evidence it really is.

A number of detailed photographs of certain portions of the lunar surface show a variety of perplexing features. Already some fantastic and downright absurd claims have been made concerning many of these, but in my opinion most of the "objects" allegedly seen are mere geological manifestations of a rather less usual nature, or are due to light and shadow effects causing the eyes to play tricks on the senses. Nevertheless it must be admitted that there *are* one or two which are less easily explained, though this does *not* necessarily imply evidence of alien visitation in the past. It may be of interest to list a few of these.

1. In part of the uplands between the Mare Crisium and the Mare Tranquillitatis a number of bridge-like objects appear to span chasms. These could be freakish geological features, though it must be admitted that the objects *are* incredibly bridge-like and show a certain rough symmetry (NASA photo number 72-H-835).

2. In the Bullialdus-Lubinicky area there is an object with a vague "vehicular" appearance. By vehicle in this context is meant not a form of spacecraft but a machine capable of traversing and transporting living creatures and equipment over the less rugged parts of the lunar surface. Chances are that the object is merely a large chunk of isolated rock which, by sheerest coincidence, has been given some of the more obvious characteristics of a "Moon-crawler" or the like. The general impression is one of a cab-like body. Indeed it bears a sort of vague resemblance to a diesel "switcher" or shunter of the kind seen in railroad freight yards. But that, at least, is surely one possibility that can be ruled out! (NASA photo number 72-H-1387).

3. On a portion of the rim of the crater King, thin tenuous dust clouds appear. Now there is no air or wind on the Moon to raise or blow dust around. They could be due to volcanic fumaroles still emitting intermittent puffs of gas. Such effects are by no means uncommon in terrestrial volcanic regions, especially in cases where volcanic activity has become merely vestigial (NASA photo number 72-H-837).

4. What look like vehicular tracks appear on part of the floor of the large crater Tycho. Thin cracks due to natural origins, perhaps seismic in character, or tracks left by a vehicle of sorts? (NASA photograph 72-H-837.)

5. On the floor of the crater Vitello there is a peculiar trail, as if some heavy (on the Moon, a very relative term!) object had been rolled or dragged along. A rational reason would be a line of small fissure volcanic conelets along a line of structural weakness. Such lines of small volcanic cones are by no means unknown on Earth. Under high magnification the appearance is hardly that of small extinct conelets. Indeed it does rather look as if "something had passed that way," but perhaps it was only a large boulder which for some reason, probably a moon quake or moon tremor, rolled along by virtue of lunar gravity.

6. Inside a rayed crater on the Oceanus Procellarum a small, longish, pale object appears. This too has vague "vehicle" characteristics (NASA photograph 67-H-327).

7. In crater Tycho appears a sort of edifice which under magnification looks not unlike some odd structure, or even a large grounded space vehicle. Once again this could be a combination of pure coincidence and/or the effect of light and shade. The stark "soot and whitewash" of the Moon's airless terrain can play some very queer tricks (NASA photograph 67-H-1651).

8. Two wheel-like objects (the "flying saucer" enthusiasts will no doubt approve and endorse this one) appear near the rims of a small crater in the Frau Mauro region. Whereas they *could* be two small craterlets, the usual effect of light and shade in craters is apparently missing (NASA photograph 70-H-1630).

9. In a region where the Oceanus Procellarum merges with the Herodotus Mountains there appear several platforms and edifices which cast clear and extremely well-defined narrow *triangular* shadows. These are really quite intriguing and are vaguely reminiscent of Aztec monuments—but we can assume that the Aztecs never quite made it to the Moon.

The above nine examples are probably the most appropriate in the circumstances, but whether or not significance should be attached to any of them remains a very moot point. Some writers have claimed evidence of actual *mining* operations taking or having recently taken place on the Moon, especially on the far side. With all due respect for the opinions of these writers, this does seem more than a little far-fetched. If it is merely minerals that are sought by aliens several light years distant, surely they could have found all they required and more on the planets. moons, and asteroids of their own system. Certainly our own evidence to date is that the Moon is probably quite well endowed with mineral wealth, especially aluminum, thorium, nickel, titanium, and even uranium. Admittedly the last could be more significant, but only just. Light years to get it is still a long way to go.

If we eliminate mining by aliens and other rather outlandish concepts, then why should representatives of an advanced alien race from the world or worlds of another star bother about the airless and inhospitable Moon? Why not just head straight for Earth which, surely to any cosmic race akin or closely akin to ourselves, must be infinitely more tempting and advantageous? As a base from which eventually to attack and take-over Earth perhaps, or at least to keep it under temporary surveillance? Such a prospect seems a little more reasonable, though not very much. Did Earth after all appear considerably less inviting and suitable at close range? Remotely feasible perhaps. Certainly when more extensive manned exploration of our satellite again gets under way, investigation of some of the above features should be attempted. After all it only requires a few hieroglyphics carved on a piece of Moon rock to answer that greatest of all questions—are we, or are we not, alone in the universe?

The same principles apply in the case of Mars, or indeed of any of the other satellites or solid planets within our Solar System. But certainly in the magnificent panoramic vistas of Mars produced by the two American Viking landers, there has appeared nothing that could even be remotely construed as an alien artifact or as evidence of past intelligent presence. Admittedly if there had been Martians and they had landed such a probe on the sandy wastes of the Sahara, or any other terrestrial desert area, they would have observed no evidence of life or intelligent presence either, unless a mirage had shown them the waters and palm trees of an oasis a few score miles away. This would surely have confused them—a case of first you see it, then you don't! If their probes had been equipped with means of radio or television reception, however, they might have got their proof in the shape of some of the rubbish which goes out over the air waves in the guise of entertainment.

In one respect so far, Mars has introduced a rather perplexing aspect in the peculiar canyon-like features which have appeared in certain of the Mars orbiter photographs. The crux of the matter is quite simply how all the material was removed from these meandering chasms. The wind—and wind storms on Mars, as we well know, can be very violent and prolonged indeed— seems the likeliest answer. Yet on reflection this hardly seems to represent a likely mechanism to remove such vast quantities of solid matter and deposit it (presumably) somewhere else. And even were this so, surely that same wind would be just as likely to deposit another load in the now vacant canyon. Now how are canyons formed on Earth? Let us take what is probably the most classic example, that of the magnificent and truly inspiring Grand Canyon in the United States, certainly one of the natural wonders of the world. We know the answer to that one well enough: by the eroding, gouging action of the Colorado River over millions of years. In doing so it has opened up what is virtually a geologist's paradise, laying bare the lithographic history of our planet in that region over a multitude of geological periods. Water created the Grand Canyon of Colorado just as it has created a number of other imposing canyons on our planet. But where on Mars was there ever water in that quantity or for such a vast period of time? At best this still seems most unlikely.

At this point it is all too easy to rush in with our alien mining friends and their equally alien bulldozers, mechanical diggers and giant automated scoops—even perhaps alien explosive charges, chemical and nuclear. Much of that equipment would certainly have been necessary to gouge out canyons up to two hundred kilometers wide, thousands of kilometers long, and up to about six kilometers deep. Why should any alien race, Martian or otherwise, have engaged in all this frenetic mining or excavation activity? Any intelligent beings that did this must have wanted something that was that far under the Martian surface very badly. Such possibilities just do not add up. They belong exclusively to science fiction.

Let us now look briefly at a possible alternative *natural* reason for these vast Martian canyon systems—that these are canyons gouged out, not by water, but by lava. We are by now well aware that Mars in its past has seen a great deal of large scale volcanic activity. The presence of the huge volcano Mons Olympus, among others, is in itself adequate proof of that. Now rivers of lava are certainly not unknown on Earth. Even lava "tunnels" are not uncommon, but lava rivers generally flow along ravines, canyons, and other forms of channels that are already present. The normal tendency for lava emitted by a conventional volcanic crater is for it to flow down the mountain side from the crater edge in the manner of water, though becoming increasing sluggish the farther it goes, since it is cooling all the time and becoming steadily more viscous. In fact, like water, it tends, on the whole, to take the line of least resistance, i.e., finding its way into gulleys and channels already present. Gouging action on a really deep and immense scale does not take place. It may be mentioned in passing that some terrestrial volcanoes produce a lava with so high a silica content (andesitic and rhyolitic lavas) that it is more likely to block the central conduit leading to the magma chamber of the volcano. When pressures mount sufficiently the volcano blows itself apart, the classic example being the volcano Mont Peleé on the island of Martinique in the West Indies which, on Ascencion Day in 1902, burst open sending out a glowing cloud of superheated gases and lava particles that entirely wiped out the thriving and lovely old city of St. Pierre. Out of 30,000 inhabitants only *two* survived!

In other types of igneous activity magma wells up through fissures and other points of weakness in the Earth's crust in the form of a vast, cooling and solidifying "lake" or plateau, e.g., that huge expanse of lava which forms the south central plain of India known as the Deccan, the great outpouring of lava which in tertiary times took place in northwestern Scotland and northeast Ireland, and responsible, among other features, for Fingal's Cave on the island of Staffa and the Giant's Causeway on the Antrim coast. Basaltic landscapes in parts of the United States are also typical of such action. We cannot therefore immediately attribute the canyons of Mars to such action, unless the nature of the Martian terrain and the vast extent of lava emission produced a combination of conditions unknown on Earth.

Another possibility is that these canyon-like features were caused by contortions of the Martian crust at an early stage in its geological history. The main objection to this premise is that they do so *look* like vast river systems on Earth, except of course that they do not, or at least do no longer, contain water. But perhaps, after all that has been said, they once *did*, indicating that, whatever our present beliefs and objections to the contrary, Mars did at one time, aeons ago, contain copious amounts of water on and below its surface and in its atmosphere. Because of the much lower Martian gravity this disappeared relatively quickly.

Only if all these natural explanations prove invalid are we entitled to ponder and give serious thought to the much more science-fiction possibility of non-terrestrial hands having been at work in past ages. But here again we must stop and think. Why aliens from the worlds of another star? Why not just Martians, real indigenous Martians, representatives of a race long extinct? Perhaps they were hardy creatures able to withstand cold.

These are very early days in the exploration of our own Solar System. The Moon will eventually be thoroughly explored, as will Mars. So presumably, given sufficient time, will the large moons of the system, i.e., those of Jupiter and Saturn. In September 1979 news was received that a probe of Titan, one of the moons of Saturn, revealed its having the type of primary nitrogen and hydrogen atmosphere once possessed by our own planet. Even Pluto, in time, will be reached, though at a distance of some five and a half light hours this is unlikely to occur soon. Venus we can certainly exclude in its present form; Mercury also because it is too hot and airless. And this goes for Jupiter, Saturn, Uranus, and Neptune as well because they are high gravity, gas giants probably devoid of a solid surface. But on these other more appropriate places, artifacts of an alien culture *might* just be found.

⑯

Galaxies Without End

IT SEEMS SOMEHOW fitting in this, the concluding chapter of a book which has dealt exclusively with the subject of planets, to carry the saga to its logical conclusion. We started with a brief and reasonably updated look at the much more familiar planets which orbit our Sun. Only then did we venture out into the great deeps of interstellar space to investigate those planets which orbit other suns. This done it might seem that we must of necessity have come to the end of the story. A little reflection will show this is not necessarily so.

The planets of our Sun and of the other appropriate stars we have been considering have one feature in common. Both belong to the same galaxy, that which we call the Milky Way. But there are, as we are now well aware, an infinite number of galaxies in the universe, just as there are a seemingly infinite number of stars in a galaxy. Let us therefore ponder very briefly on the theme of planets orbiting stars within *other* galaxies.

It must be obvious that, whereas we are able to search for planets of other stars in our *own* galaxy, we cannot possibly do this in respect to their opposite numbers in *other* galaxies. In this instance we can think only in the most general of terms.

If it is typical for a high proportion of stars in our galaxy to possess planets, many perhaps habitable, then it is eminently reasonable to accord the same distinction to other galaxies. There seems no valid reason why ours should be unique in this respect. We are thus presented with the picture, not only of a galaxy containing an immense number of planets, but of the entire universe and an *infinity* of worlds. It could just also mean life and civilizations without end. And that by any standard is a sobering and an intriguing thought!

So let us therefore dwell briefly on those other galaxies, always remembering as we do that in our context we are not just considering them as other galaxies, not just other great assemblages of stars, but also as regions in space where immense numbers of other worlds exist. As we advance in thought and by telescope ever more deeply into metagalactic space, more and yet more of these great and majestic "island universes" and the myriad of worlds they contain are to be found. But these are worlds so incredibly remote that they make even those on the far side of the Milky Way our virtual near neighbors.

The fact that the Milky Way is a galaxy of stars of which our Sun is merely a single, undistinguished member was not established until the advent of the telescope, though the possibility had been foreseen by some astronomers years earlier. To most, however, it was a mysterious, lovely, diaphanous band of light girdling the night sky, hence the name of Milky Way. It was not until 1924 that it was confirmed that the great Andromeda nebula, M31, was in fact *another* great galaxy of stars and not merely a cloud of elemental gas and dust *within* the confines of the Milky Way. The position prior to then is clearly epitomized by the late Sir Robert Ball who, in his *Story of the Heavens* (1894), said briefly of the spiral nebula, "It is believed that some of these nebulae are sunk in space to such an appalling distance that their light takes centuries to reach the Earth."

It was the late, and now almost legendary Edwin Hubble who grasped the true situation. Unlike Ball, Hubble realized that these objects were not just *hundreds* of light years away but were, in fact, located at distances greatly in excess of objects within the galaxy. Using the 100-inch Hooker telescope of the Mount Wilson Observatory, Hubble simply revolutionized earlier concepts of the universe by showing unequivocally that stars are grouped in aggregations or galaxies of which our own, the Milky Way, is merely one of innumerable galaxies, most being so remote as to appear only as small diffuse nebulae. By so doing Hubble brought an end to a lively and sometimes highly acrimorious debate, since many astronomers had resolutely maintained that *all* nebulae were members of the Milky Way system. In the case of true nebulae such as M42 in Orion, this is indeed the case. This had been clearly demonstrated by the spectroscope that had indicated that certain nebulae lying close to the plane of the Milky Way are simply luminous clouds of gas. At the same time the spectroscope also showed that the spectra of many other seeming clouds of gas bore a most marked resemblance to *stellar* spectra. It was into the second category that, for example, the Andromeda "nebula" fell. This object was, in fact, an external galaxy of stars, an extragalactic nebula though it must be emphasized that in this context the term nebula does *not* bear its original connotation of gas cloud. A term used to describe such objects a few decades back was "island universe" though nowadays, for some reason or other, it seems to have fallen into disfavor.

Cepheid variable stars were soon spotted among the brightest stars of the Andromeda galaxy, and since the intrinsic luminosities of these bodies are known with reasonable certainty, it became possible to estimate how far distant they (and thus the galaxy containing them) happen to lie. At the time this worked out at 1.5 million light years. And this, let us hasten to remind the reader, is the *nearest* other galaxy of stars (and of planets) to our own. Overall dimensions of this galaxy are comparable to those of our own, though that in Andromeda is presently regarded as being the greater. Rotational velocities and densities are similar.

At this point it is also interesting to draw attention to the fact that the Andromeda galaxy is the most remote celestial object which it is possible to discern with the unaided eye. Any decent star atlas will pin-point its position among the stars of the constellation Andromeda. It can be located fairly easily as a small, fuzzy patch of faint light any clear and moonless night during that part of the year when the constellation is visible in the night sky, notably during late autumn and winter. It is important to recall, as one views this faint diffused patch of light, that it represents an *entire galaxy,* one containing even more stars than our own, and presumably also more planets. A small telescope or pair of binoculars shows the object as a bright, roughly oval mass but it requires a telescope of considerable aperture to reveal anything of its true spiral nature.

The advent of the great Hale 200-inch telescope of the Mount Palomar Observatory shortly after the close of World War II rendered it possible to photograph galaxies not just a few million light years distant, but many as remote as two thousand million light years. These are distances so vast and awesome that the human mind is totally incapable of coping with them. Here are entire galaxies like our own great Milky Way, the light from which, travelling at over 186,000 miles per second, has taken all of 2,000 million years to reach us. Clearly there can be no hope whatsoever of our identifying planets surrounding stars in these incredibly remote galaxies. Indeed there is little hope of discerning even the individual stars. Yet stars there must be. And, if our previous reasoning is sound, planets there must also be—and very likely life also. The light reaching us today from these galaxies began its journey towards us when the first elementary forms of life were only just beginning to appear on Earth.

More recently, clusters of galaxies in the constellations Hydra and Coma Berenices were photographed using the Palomar 200-inch reflector. These are reckoned to be as much as 5,000 million light years remote. Since this is more or less the present age of the Earth, we must accept that light from these galaxies began its journey toward us as Earth was being created, or perhaps even before that. Thus if intelligent or potentially intelligent life was already then in being on worlds within these galaxies, we must ponder over what stage of technological advancement it has now reached, or whether it long ago blew itself and its works into nuclear oblivion. We may ponder. We can never know.

Distances of the order of 5,000 million light years may represent the ultimate for optical telescopes, or at least so far as those of the present generation working from the surface of the Earth are concerned. The coming of radio astronomy and of radio telescopes has led to the discovery of whole galaxies at much greater distances. Here indeed are masses of suns (and planets), assuming they still exist, which came into being long before the creation of our Earth and Sun. We have no reason to suppose that there are not hosts of others even farther outlying well beyond the discerning power of contemporary radio telescopes.

Surely at this point we should be able to cry enough! This may not be so. The possible existence of black holes, those strange regions of space where a dying star of appropriate dimensions has crushed itself out of existence could, it is postulated by some cosmologists, lead from this universe into another unknown one in a dimension beyond our understanding. If this is so then in this other universe too there will be an infinity of galaxies of stars and planets—and perhaps therein black holes leading to yet further universes, further galaxies and further planets. Does it all have a meaning or is it all the result of a great cosmic accident? We will never know.

Appendix 1

TALK OF REACHING the planets of other stars may now seem purely academic. Nevertheless it is a measure of man's progress and ambitions in space that only twenty-two years after the first artificial satellite went into orbit, competent, qualified space technologists are already giving the matter serious thought. This certainly does not mean that package tours to Alpha Centauri or around the worlds of the Barnard Solar System are imminent. What it does mean however, is that interstellar travel is becoming less and less the prerogative of science fiction. In a sense we are seeing a repeat run, on a much more protracted time scale, of the events which led to *interplanetary* travel.

Interplanetary travel had a simple beginning—merely a small sphere with protruding antennae that went bleeping around Earth on October 4, 1957. But that simple beginning led directly to Armstrong and Aldrin putting mankind's footsteps on the Moon twelve years later, excellent landscape photographs direct from Mars in 1976, and close-up shots of the moons of Jupiter in 1979. Great things have small beginnings. So also will it be for interstellar travel. Clearly our opening gambit will not be a headlong rush to Alpha Centauri. It will be, indeed must be, a precursory mission not too deep into interstellar space.

A highly competent and detailed study of such a tentative interstellar probe was recently carried out by a number of scientists at the Jet Propulsion Laboratory, California Institute of Technology in Pasadena. (L. D. Jaffe, C. Ivie, J. S. Lewis, R. Lynos, H. N. Norton, J. W. Stearns, L. D. Stimpson and P. Weesman.) Readers interested in the full scope of this study will find it in full in the journal of the British Interplanetary Society, vol. 33, pp. 3–26, 1980. In this appendix only the basic essentials will be outlined.

The overall objective of the study was to establish certain scientific aims, techniques, and technology requirements for an unmanned mission some distance into interstellar space. Bizarre concepts such as the peculiar relativistic effects due to near-optical velocities are largely ignored since the attainment of such velocities must lie in a much more distant future.

Despite opinions to the contrary held by certain members of the project team, it was deemed inappropriate to base the study on the premise of a mission to an actual star. This decision was based on the fact that the hyperbolic velocity attainable to escape from the gravitational pull of the Sun and its retinue of planets during the period under consideration (the years 1985–2015) would probably be of the order of 100km/sec (22,500 miles/hour). The nearest star being 4.3 light years distant (4.10^{13} Kms or 25 million million miles) a space vehicle with this velocity would take 10,000 years to reach it. This seemed not only grossly unrealistic but undesirable for two quite specific reasons:

a. The production of a space vehicle having a designed life span of 100,000 years by the end of the present century is simply not feasible.
b. Propulsion and velocity capability will almost certainly increase with time. Doubling the velocity presently envisaged should take about twenty-five years. The mission time would thus be halved (5,000 years). Consequently a space craft launched later could conceivably arrive *before* the first one!

Seven specific objectives are enumerated in the study. Of these, four are regarded as primary and three as secondary.

1. **Primary**
   a. Determination of the characteristics of the interstellar medium
   b. Assessment of the characteristics of the heliopause (i.e., where solar wind meets incoming interstellar medium)
   c. Precise determination of stellar and galactic distances by virtue of distance measurements to certain of the "nearer" stars
   d. Determination of characteristics of the Solar System as an entity (e.g., total mass, interplanetary gas distribution, etc.)
2. **Secondary**
   a. A close look at Pluto and its satellite(s), and rings if any—assuming that by then this has not already been accomplished

b. Determination of the characteristics of remote galactic and extra-galactic objects (gaseous nebulae, star clusters and spiral galaxies).
c. Evaluation of the problems likely in making scientific observation of another solar system from a space vehicle: This could be achieved by giving the mission a "return" potential so that the Solar System could play the role of objective in a simulated stellar encounter. As the spacecraft approached the Solar System it would be possible to establish fairly precisely at what distance planets, (and later) satellites, asteroids, and comets became discernible. If the mission were merely one way (i.e., outward from the Solar System only) the same data might be obtained but in reverse order—at what distance did asteroids, moons, etc., and later planets become lost to sight.

The mission is seen as one extending 400 to 1,000 astronomical units from the Sun (37 thousand million miles to 93 thousand million miles). The study also embraces appropriate space craft systems and technology requirements.

If something along these lines can be accomplished around the turn of the century (and there seems no reason why it should not) then a very real start will have been made toward the eventual realization of interstellar travel. Out across immensity to the stars: this has long been regarded by many as man's final and irrevocable destiny. The opening gambit just outlined may not seem particularly ambitious but it could lead directly and in time to an infinitely exciting future!

But when, the reader may well be tempted to ask, will that infinitely exciting future really begin? When will man himself first venture out into the great interstellar abyss? This is clearly a very difficult question to answer. Any estimate, even one of the most reasoned kind, tends to be something of a "guesstimate." This is hardly surprising in view of the many imponderables involved. We would not wish to subscribe to pessimistic predictions. At the same time no useful purpose whatever can be served by those of an overtly optimistic nature. Somewhere between the two lies truth. We would assume that any impending disaster to the Sun likely to take place within a generation or two would serve as a very powerful motive force indeed. For example, but for the advent of World War II, it is almost certain that the harnessing of the power of the atom would not have taken place for several decades after it did. War and impending disaster concentrate the mind most wonderfully! They even succeed in getting politicians to move and make some right decisions for a change. However, unless astrophysicists are hopelessly wrong (which seems unlikely) we have 5,000 million years before the Sun swells out into a red giant star and consumes our planet. Instability in the Sun or extensive solar flaring would seem to represent a slightly earlier menace.

However, assuming that our Sun is going to behave itself and offer no threat in the foreseeable future, the advent of interstellar travel is not going to be accelerated in an undue way. That could be just as well, as haste can lead to risks and risks to disaster. Where then do we stand with respect to a reasonable estimate of the date of the first *manned* interstellar mission? By this we mean a true *colonization* mission and not just a brief trip into interstellar space before a return to the relative security of a base, say on Pluto.

Much has been written and many predictions made, and in attempting to find something that seems like a reasonable estimate in view of all the problems relating to time, distance, and propulsion, the writer has waded through a considerable number of papers on the theme appearing in erudite and thoroughly respectable scientific journals. As might be expected such estimates vary greatly. However, what seems to be one of the most reasonable and best reasoned is of fairly recent vintage. It is contained in a paper by Gregory L. Matloff of the New York University Department of Applied Sciences and Eugene F. Mallove of Interstellar Enterprises, Holliston, Massachusetts, bearing the title of "The First Interstellar Colonization Mission" (Journal, British Interplanetary Society, vol. 33 pp. 84-88, 1980). Matloff and Mallove believe the first interstellar colonization mission may be launched within the next two hundred years using pulsed thermonuclear propulsion in a star ship having a mass around 2×10^{10} gms (20 million kilograms). Initially this star ship would have a complement of from fifty to a hundred persons, and would be launched toward an extra-solar planetary system containing a terrestrial type planet. Propulsion would be achieved by staged fusion micro-explosions of lithium-hydrogen or boron-hydrogen. This mission would, of course, have to be of the "generation" or "space-ark" type. Neither writer makes light of the difficulties. There is, however, one snag which is not really mentioned. Suppose the destination planet is reached and found to be inhabited? This represents one of these inevitable questions, and here too there are wide variations in opinion. Among the most recent is that contained in a well thought-out paper by Alan Bond and Anthony R. Martin of the Culham Laboratory, Oxford, England, entitled "A conservative Estimate of the Number of Habitable Planets in the Galaxies" (Journal, British Interplanetary Society, vol. 33 pp. 101-106, 1980). The estimate arrived at is that our galaxy contains 5.3 million habitable planets having a *mean* separation of 140 light years and 2.4 million planets orbiting stars of sufficient age for intelligent life to have formed, mean separation in this case being 180 light years. Both Bond and Martin very sensibly point out that their paper does *not* estimate the *probability* of such life forming. In interstellar space man will simply have to go forth and take his chances. No doubt he will, just as he has done in other great spheres of human endeavor. If the Sun should be proving increasingly hostile, for whatever reason, he will have little choice. It is as simple as that.

Appendix 2

I N REFERRING TO what tricks the Sun might get up to we have had to be careful not to transgress the established laws of astrophysics by implying that the Sun could reasonably go on the rampage the way certain other stars can and sometimes do. At the same time it would be wrong to give the impression that the Sun is a totally docile star which never strays from the straight and narrow path.

Fairly recently reports have been appearing in a number of scientific journals concerning the behavioral aspects of the Sun. These tend to make interesting reading and to give reason for a measure of reflection. The fact is that there is accumulating evidence that the Sun is (and has long been) a variable star. As long as man was pinned to the Earth his sources of data were inevitably restricted. Now all this has changed. As a consequence of measurements made from spacecraft, Mariner and Viking photographs of Mars, and lunar rock samples to name but three, vital new evidence relating to solar behavior has been and is being obtained. To this we can add evidence from purely terrestrial sources, e.g., pollen counts, tree rings, deep sea sediments, glacial deposits, and climate records.

When all this data is integrated and collated certain fairly definite facts begin to emerge. It would appear that the Sun displays variability based on three distinct time scales:

a. 0–100 years
b. 100–100,000 years
c. 100,000 to 1,000 million years.

Short term variability (0–100 years) is manifested in the contemporary Sun by virtue of solar flares and sunspot activity, features which we have already touched upon. Extremely short term variability (hours to days) is shown by changes in ultraviolet and X-ray output. Variation in the composition of solar flare particles has been measured by spacecraft while examination of lunar samples indicate quite significant variations in proton flux of the order of decades. The effect of the eleven-year cycle of solar activity on the flow of incoming cosmic rays from galactic sources has been known for some time. However recent measurement of cosmic-induced radio-activity in meteorites falling during the years 1967–68 indicates the effect on this flow to be about *three times* greater than anticipated.

In the intermediate variability term of 100 to 100,000 years, extensive use must, of necessity, be made of historical records. Unfortunately not all of these are paragons of accuracy, and we must be very careful before drawing conclusions from this source. However the stamp of the Sun's activity has been etched onto certain natural materials, e.g., pine tree rings and iron meteorites. It would appear that there was apparently a period of greatly reduced solar activity and an almost complete absence of sunspots between the years 1645 and 1715.

Solar flare intensity appears to have been constant for a period of around 500 thousand years, though the possibility of greater flare activity is being seriously considered in respect of a period around ten thousand years ago. This belief is backed up by the excess amount of radioactive carbon 14 found in lunar surface rocks and soils.

Really long term variability (100,000 to 1,000 million years) demands a knowledge of what the recently-born Sun must have been like. Up to about ten years ago this was very much the province of the theorists, but now scientists are extracting valuable information from solar particles contained in certain rocks (breccias) from the lunar highlands that were exposed to rays from the very young Sun. Flare activity seems to have been fairly intense around 4,200 million years ago, leaving particle tracks in meteorite grains exposed to the Sun at that time.

Compared to the variable stars we considered in chapter 8 there is a reassuring staid constancy about the Sun. Nevertheless there *is* some evidence of variations and this currently appears to be mounting. Perhaps, and this seems quite likely, there is no such thing as a truly non-variable star. All stars, by their very nature are probably variable to a greater or lesser degree. Those distant stars which we regard as non-variable might show their true colors were we able to observe them at closer range. There are exceptions even to this. The lovely red giant star Betelgeuse in the right "shoulder" of Orion shows a limited but distinct variability—and Betelgeuse lies some 500 light years distant!

At present our Sun appears to give only a very occasional hiccup. We trust it will be a long time indeed before it decides to give a cough!

Bibliography

1. The Sun—Our Lifeline

Clark, David. "Our Inconstant Sun." *New Scientist* [London], 18 January 1979, pp. 168-79.

Dicke, R. H. "Is There a Chronometer Hidden Deep in the Sun?" *Nature* [London], 14 December 1978, pp. 676-80.

New Scientist. "How Hot Supernovae Could Freeze Up the Earth," 9 February 1978, p. 363.

———. "Comets—Harbingers of the Freeze-up?" 13 April 1978, p. 85.

———. "Scientists Shrink from Smaller Sun," 21 June 1978, p. 982.

———. "Solar System on Collision Course," 27 July 1978, p. 271.

Walgate, Robert. "The Shivering Sun." *New Scientist,* 25 May 1978, p. 509.

Weyman, R. J. "Stellar Winds—Stars that Eject Matter into Space." *Scientific American,* August 1978, pp. 34-43.

Wolfendale, Arnold. "Cosmic Rays and Ancient Catastrolphes." *New Scientist,* 31 August 1978, pp. 634-36.

2. Our Sister Worlds

Burgess, Eric. "Probing the Clouds of Venus." *New Scientist,* 7 December 1978, pp. 763-65.

Harrington, Robert S. and Betty J. "The Discovery of Pluto's Moon." *Mercury* [San Francisco], January-February 1979, pp. 1-3.

Hartmann, William. "The Watery Past of Mars." *New Scientist,* 28 June 1979, pp. 1083-85.

Heath, Martin. "Last Outpost of the Solar System." *Ad Astra* [London] 3:1 (1978).

Hindley, K. "Chiron—the Celestial Centaur." *New Scientist,* 2 February 1978, pp. 300-301.

Hunt, G. E. "A Pioneer's View of Uranus." *Nature,* 26 April 1979, pp. 777-78.

Hunter, D. M. "New Surprises from Uranus." *Nature,* 2 November 1978, p. 16.

Lawton, A. T. "Charon—A Companion to Pluto." *Spaceflight* [London] 20:12 (1978), pp. 428-29.

Lewis, Richard. "Yes, There Is Life on Mars." *New Scientist,* 12 October 1978, pp. 106-108.

Morrison, Nancy D. "Volcanoes on Io; A Ring Around Jupiter." *Mercury,* May-June 1979, pp. 63-66.

New Scientist. "Radar Finds a Big Volcano on Venus," 26 January 1978, p. 215.

―――. "Faint, but Red as Amalthea," 16 March 1978, p. 729.

―――. "Martian Surface in Good Spirits," 6 July 1978, p. 19.

―――. "Groove Network on Phobos Bears Witness to Shattering Blow," 6 July 1978, p. 340.

―――. "Odd Polarization Hints at Icy Craters on Jovian Satellites," 3 August 1978, p. 340.

―――. "Mars Seems Quite Dead," 31 August 1978, p. 621.

―――. "Uranus's Luck Does Not Hold Out All Round," 23 November 1978, p. 607.

―――. "More Satellites for Saturn?" 4 January 1979, p. 22.

―――. "Satellites Hold Rings Around Uranus," 11 January 1979, p. 92.

―――. "Neptune Is More Way Out than Pluto," 18 January 1979, p. 157.

―――. "Pioneer Finds Grand Canyon on Venus," 15 February 1979, p. 462.

―――. "Voyager Spots Volcanoes on Jupiter's Moon," 15 March 1979, p. 844.

Ridpath, Ian. "Pluto Looks Smaller than Ever." *New Scientist,* 27 July 1978, p. 273.

Sutton, Christine. "Jupiter's Enigmatic Variations." *New Scientist,* 5 April 1979, pp. 21-23.

―――. "A Close-up Look at the Solar System's Largest Planet." *New Scientist,* 19 July 1979, pp. 217-20.

Trimble, Virginia. "It's a Nice Planet to Visit but I Wouldn't Want to Live There." *Cosmic Search* [Delaware, Ohio] 2:1 (1980), pp. 13-16.

3. Planet 10

Lawton, A. T. "Many Shades of the 10th Planet." *Spaceflight* 21:3 (1979), pp. 115-23.

4. In the Beginning

Brecher, K. et al. "Is There a Ring Around the Sun?" *Nature,* 1 November 1979, pp. 50-52.
Gwynne, Peter. "Venus Probes Solar System Birth." *New Scientist,* 21/28 December 1978, p. 916.
Herbst, William and Assousa, G. E. "Supernovae and Star Formation." *Scientific American,* August 1979, pp. 122-29.
New Scientist. "Astronomers Look Out for Planet Birth," 19 January 1978, p. 157.
Paterson, David. "A Supernova Trigger for the Solar System." *New Scientist,* 11 May 1978, pp. 361-63.
Reeves, H. "Origin of Solar System." *Mercury* 6:2 (1977), pp. 7-14.
Schramm, D. N. and Clayton, R. N. "Did a Supernova Trigger the Formation of the Solar System?" *Scientific American,* October 1978, pp. 98-113.

5. Piercing the Veil

Fennelly, A. T. et al. "Photometric Detection of Extra-Solar Planets Using L.S.T.-type Telescope." *Journal of the British Interplanetary Society* [hereafter *JBIS*] 28:6 (1975), pp. 399-404.
Herbig, George H. "A Universe Teeming with Planetary Systems." *Mercury* 5:2 (1976).
Lawton, A. T. "Photometric Observation of Planets at Interstellar Distances." *JBIS* 12:8 (1974), pp. 365-73.
Martin, A. B. "Detection of Extra-Solar Interplanetary Systems II." *JBIS* 27:4 (1974), pp. 881-906.
_____. "Detection of Extra-Solar Interplanetary Systems III." *JBIS* 28:3 (1975), pp. 182-90.
New Scientist. "Space Telescope Could Look for New Planets," 2 November 1978, p. 356.

_____. "Pulsar with a Planet," 22 November 1979, p. 607.
Richards, G. R. "Planetary Detection from Interstellar Probes."
  JBIS 28:8 (1978), pp. 579-85.
Spaceflight. "Project Orion: A Method for Detecting Extra-Solar
  Planets," 19:3 (1977), pp. 90-92.

6. We Are Not Alone

Lawton, A. T. "Nearest Other Solar System." *Spaceflight* 12:4
  (1970), pp. 170-73.
New Scientist. "Space Is Lonelier than We Thought," 15 March
  1979, p. 864.

7. Multiple Suns

Hack, Margherita et al. "Observations of the Eclipsing Binary
  Epsilon Aurigae." *Nature,* 23 November 1978, pp. 376-78.
Harrington, Robert S. and Betty J. "Can We Find a Place to Live
  Near a Multiple Sun?" *Mercury,* March-April 1978, pp. 34-37.
New Scientist. "Double Star with Planets," 22 September 1977, p.
  727.
_____. "Solar Companion Is Still Possible," 26 October 1978, p.
  277.
_____. "Gravity Waves Are Running Down a Binary," 21/28
  December 1978, p. 916.
Trimble, Virginia. "How to Survive the Cataclysmic Binaries."
  Mercury, January-February 1980, pp. 8-12.

8. Uncertain Suns

Glasby, J. S. *Variable Stars.* London: Constable & Co. Ltd., 1968.
_____. *Dwarf Novae.* London: Constable & Co. Ltd., 1970.
Irwin, J. B. "The Case of the Degenerate Dwarf." *Mercury,*
  November-December 1978, pp. 125-27.

Morrison, Nancy D. "The Mysterious Object Eta Carinae." *Mercury,* January-February 1980, pp. 12-13.

New Scientist. "Standard Star May Vary," 5 April 1979, p. 27.

_____. "Astronomers Respond Slowly to Variable Star," 28 June 1979, p. 1089.

9. Cosmic Wanderers

Hindley, K. "Did Asteroid Impacts Farther Earth's Ocean Basins?" *New Scientist,* 19 January 1978, p. 159.

Hughes, D. W. "Earth's Cratering Rate." *Nature,* 6 September 1979, p. 11.

Lawton, A. T. "Stray Planets." *Spaceflight* 16:5 (1974), pp. 188-89.

Macvey, John W. "Wandering Worlds." *United States Air Force Digest,* June 1962, p. 68.

New Scientist. "Phobos Is Probably a Captured Asteroid," 9 February 1978, p. 364.

_____. "Double Collision with Asteroids Averted," 6 April 1978, p. 21.

_____. "Groove Network on Phobos Bears Witness to Shattering Blow," 6 July 1978, p. 20.

_____. "Asteroid Gaps Explained—but Only Mathematically," 13 July 1978, p. 109.

_____. "Comets Tell of Planet that Exploded into Asteroids," 4 January 1979, p. 22.

_____. "Chiron Is Not Here to Stay," 29 March 1979, p. 1032.

_____. "Minor Planets May Have Company," 5 April 1979, p. 26.

Wetherill, g. W. "Apollo Objects." *Scientific American,* March 1979, pp. 38-49.

10. Air to Breathe

Bailey, K. "Volcanic Activity: The Continental Degassing of Earth." *New Scientist,* 1 February 1979, pp. 313-15.

Handley, K. "Earth's Atmosphere—A Lovely Fluke." *New Scientist,* 8 June 1978, p. 671.

Sellers, Ann H. "The Evolution of the Earth's Atmosphere." *New Scientist,* 26 October 1978, pp. 287-89.

11. And Lo, There Was Life

Bell, Trudy E. "History of the Idea of Terrestrial Life." *Cosmic Search* 2:1 (1980), pp. 2-10.
Briggs, M. H. "Terrestrial and Extra-Terrestrial Life." *Spaceflight* 2:4 (1959), p. 120.
──────. "Automated Life Detection Devices," *Spaceflight* 5:4 (1963), pp. 128-33.
Browne, M. W. "Life May Exist Only on Earth, Study Says." *New York Times,* 24 April 1979.
Burnell, Susan J. B. "Little Green Men, White Dwarfs or Pulsars?" *Cosmic Search* 1:1 (1979), pp. 17-21.
Cottey, Alan. "Advanced Life in the Universe." *New Scientist,* 27 April 1978, pp. 236-37.
Ehricke, K. G. "Astrogenic Environments." *Spaceflight* 14:1 (1972), pp. 2-14.
Kroto, H. "Chemistry Between the Stars." *New Scientist,* 10 August 1978, pp. 400-403.
Molton, P. M. "Is Anyone Out There?" *Spaceflight* 15:7 (1973), pp. 246-51.
Muller, H. J. "Life from Elsewhere." *Spaceflight* 5:3 (1963), pp. 74-75.
New Scientist. "We Are Not Alone," 28 June 1979, p. 1090.
Ridpath, Ian M. "Solar Systems and Life." *Spaceflight* 17:8/9 (1975), pp. 323-27.
Sagan, Carl. "The Quest for Extra-Terrestrial Intelligence." *Cosmic Search* 1:2 (1979), pp. 2-9.
Sagan, Carl and Drake, Frank. "The Search for Extra-Terrestrial Intelligence." *Scientific American,* May 1974, pp. 80-89.
Slater, A. E. "Life in the Universe." *Spaceflight* 5:6 (1963), pp. 198-200.
Spaill, N. J. "Physical Appearance of Intelligent Aliens." *JBIS* 32:3 (1979), pp. 97-102.
Tartar, Jill et al. "Searching for Extra-Terrestrial Intelligence: The Ultimate Exploration." *Mercury,* July-August 1977, pp. 3-7.

12. Per Ardua Ad Astra

Davies, P. C. W. "Anti-matter from Space." *Nature,* 8 November 1979, p. 130.

Martin, A. R. and Bond, A. "Nuclear Pulse Propulsion." *JBIS* 32:8 (1979), pp. 283–310.

Matloff, G. L. "The Interstellar Ramjet Acceleration Runway." *JBIS* 32:6 (1979), pp. 219–20.

New Scientist. "Moving Lines Hint at Superlight Speeds," 18 January 1979, p. 173.

Waldron, R. A. "The Ballistic Theory of Light and its Implications for Space Travel." *JBIS* 32:3 (1979), pp. 95–98.

Winterberg, F. "Rocket Propulsion by Nuclear Microexplosions and the Interstellar Paradox." *JBIS* 32:11 (1979), pp. 403–409.

13. Which Way?

Orme, A. W. "The Direction of Interstellar Exploration." *Spaceflight* 21:10 (1979), pp. 397–401.

Strong, J. *Flight to the Stars.* London: Temple Press Books, 1965.

14. Cities in Space

Bernal, J. D. *The World, the Flesh, and the Devil.* London: Jonathan Cape Ltd., 1970.

Clarke, Arthur C. *The Exploration of Space.* London: Temple Press Books, 1951.

Hassall, Tim. "Towards Cities in Space." *Spaceflight* 21:12 (1979), pp. 504–11.

Macvey, John W. *Journey to Alpha Centauri.* New York: Macmillan, Inc., 1965.

O'Neill, G. K. "A Lagrangian Community." *Nature, 250* 1974, p. 636.

———. "The Colonization of Space." *Physics Today,* September 1974, p. 32.

———. *High Frontier Human Colonies in Space.* London: Jonathan Cape Ltd., 1977.

Shepherd, L. R. *Realities of Space Travel.* London: Putnam & Co. Ltd., 1957.

Tsiolkovsky, K. E. *Investigation of World Spaces by Reactive Vehicles.* Moscow: Mir Publishers, (date unknown).

Index